# Lighting Design

## 照明设计

建筑·景观·艺术
Building & Cityscape & Art

凤凰空间·上海 编

江苏人民出版社

# Preface
# 序言

Before people have defined the term of design, it was only known the core motivation is for satisfying human need. In recent decades, design has been defined as a process of observing human behavior and experiences, understanding the needs, and finally the problem solving. By applying the concept of design into lighting design, we can say that the lighting design came into existence because of the desire of light. The history of light can be traced back to the natural lighting from the fire to the creation of light bulb by Thomas Edison. This development demonstrates that light is always an essential element which is as significant as the air for humanity. As the progression in technology, culture, and economy, people nowadays yearn to improve life with a better quality of light. This motivation not only changes the way people live, but also lighting design, which gets a great advancement, and we believe that lighting design is no longer a single subject.

As other fields of design, lighting design has combined with the idea of user-centred design recently. This collaboration makes design idea more thoughtful and suitable for people. When mentioning the terms of "User-Centred Design" or "Universal Design", they were originally used for industrial design which emphasised the concerning of the need from the end-users. Within a few decades, its concept has been applied to various fields, not only in design, but other non-design areas. It has become a common attitude that is conventional in most public.

Then, why do we put so much concern on lighting effect? The reason is its unique feature. Lighting is an important medium for human to perceive their environment. It not only decided whether people see an object, also how they feel about it. That is, the lighting can affect people both physically and mentally. By combining light and user-centred design, we create a lighting environment that focuses on how people feel and interact with the light. How can we make people feel better in the environment by changing the lighting effect? This is the premier consideration when developing a concept. This consideration represents that lumen is no longer the only way to evaluate lighting appropriation in a space, which means the designers should focus on the quality of light rather than the quantity, and the idea of light must be more delicate and suitable for daily life. To create environment that suits all kinds of occasions in life, for example, the illumination that helps concentration while reading, accent lighting in museum, romantic ambient light in restaurant, or a grand building at night, lighting designers gather their thoughts and feelings about the space based on their experience in light application and communicate with spatial designers, architectures, and clients continuously.

In the past few years, people have started to realise the importance of illumination for urban view and living style. They have paid more attention on lighting design, especially in Asia, such as China and South Korea, which have the maturity of LED technology in these countries. However, the fast growing in hardware also brings to light the issues, the shortage in lighting professional and education for lighting design. Lighting design can be treated as an art creation that merges both aesthetics and technology. It aims to create a harmony and assists to present the image for a space, a building, or urban landscape at day and night. Designers must have knowledge about light and understand the relationship between the light and the space in order to make an efficient lighting effect to show its value.

Regarding to the experiences, we found that lighting design is still a newborn industry in Asia. When reviewing the current situation, it is not surprising to find that most lighting designers come from different professions but lighting. Lighting design needs to take complex considerations, such as the aesthetics, space, lighting features, moreover, some engineer problems, into account at once when developing an idea and we still don't have mature educational structures and platforms which combine multiple disciplines for educating professionals in this field. Fortunately, the lighting industry has received more attention because of the anticipation on the issues of energy conservation. This fact has encouraged the collaborations of professions from government, education, research institutes, and real industry to build a credentials and comprehensive educational environment which integrated all of essential knowledge for the lighting design. From the perspective of lighting designers, we are glad to see this change since it will doubtlessly promote a great progress in industry of lighting design in future Asia.

GUANG Architecture Lighting Design
Alexander, C.N. Sun / President & Design Principal
Michelle, T.Y. Liu / Designer

我们想"设计",最终是为问题提供一种解决的方法,谈及"设计"行为最核心的动力,是为满足人类的需求而产生的。同样的概念在照明设计的领域来说,所表示的也是人类因对"光"的需求而造就了照明设计的存在。在"光"的历史中,从最原始的火光到之后爱迪生发明了灯泡,显现出对人类而言,光与空气是同等重要的生命元素。发展至今,因社会、经济、整体环境的进步,人类逐渐地想要藉由设计的行为让光环境达到更进一步的提升,并且设计概念的中心思考是以"人本为主"。"User-Centred Design"、"Universal Design"等词汇原用于工业设计针对产品及服务,强调以最终使用者的需求出发考量,是设计的一个过程。直至今日,"User-Centred Design"的概念已被广泛地应用在不同领域,不仅是设计工作,更是普遍的态度与价值认同。

因此,回到照明设计,光的运用不再只是专注在亮(量)即可,而是应以更为细腻且生活化的方法来思考——"如何让环境因为光而让人感觉更好?"人类因光而能够看见、了解事物的样貌,也因此照明所能够影响的,不只是空间的明亮度,同时更直接地影响了观者的心理。能提高专注力的灯光、博物馆重点式照明、具有浪漫气氛的餐厅、壮观的建筑外部结构等,照明设计师凭借着使用"光"的经验,与空间设计师、建筑师、业主不断重复地交流及整合对环境的感受,去创造出一个最为适合的环境氛围。

在亚洲地区如中国、韩国的照明设计因LED电子工程技术集中成熟,照明设计所能够发挥的机会越来越多,同时,社会也逐渐意识到照明对都市景观、生活环境的重要性,但硬件的快速发展也暴露了照明专业人才及设计师美学素养缺乏的问题。照明设计是一门集结了人文及科技的艺术,演绎了空间、建筑、景观在日夜间调和的美感,设计者必须对光的性质和空间的关系具有足够的了解和知识,才能让其中的各项元素发挥其影响及展现其价值。

在过去的经验中我们发现的是,因照明设计需考量的元素较为复杂且跨学科领域,再加上目前学校的教育尚未有完整的课程架构能提供足够学养的环境,至今照明设计依然普遍被视为新兴。且照明设计师需要是跨领域的工作者,人才素养的培育更必须结合实际产业的环境作为教育的平台。庆幸的是,照明产业的动向也因全球节能议题的重视,渐渐地在政府、学校、研究机构、产业界专家合作之下,结合各领域与照明相关的必要专业知识,开始推动一系列能力认证及教育学程的建立。就我们照明设计师的角度来说,特别对亚洲区的照明整体产业而言,这无疑地是一个具有前瞻性及计划性的进步开端。

光拓彩通照明设计 孙启能 / 总经理及主持设计师
刘子银 / 设计师

# Contents
# 目录

# Preface 序言
# Overview 综述

## 01 Office & Cultural 办公+文化

| | | | | |
|---|---|---|---|---|
| 038 | TBA Tower<br>夜晚中的生命质感—TBA环球经贸中心 | | 084 | "Lightstream" Dallas Convention Center<br>"光流"——达拉斯会展中心 |
| 050 | 155 North Wacker Drive<br>高性能，低维护—芝加哥北威克街155号 | | 090 | Roca Barcelona Gallery<br>黄昏里的"面纱"——乐家巴塞罗那展厅 |
| 060 | Locarno: The Ideal City<br>哲学与三色光源——洛迦诺：理想城 | | 094 | Pio Palace in Carpi<br>与建筑相融合——卡尔皮皮奥宫殿 |
| 070 | Palazzo d'Arnolfo<br>多色调白光的精彩演绎——阿诺尔夫宫殿 | | 098 | Klimahaus 8° Ost, Bremerhaven<br>LED点阵灯的演绎——不来梅东经八度气候馆 |
| 076 | Old DC Courthouse<br>照明之于国家地标的修复——华盛顿哥伦比亚特区法院 | | 102 | Fondazione Arnaldo Pomodoro, Milan<br>光与艺术品的对话——米兰阿纳尔多·波莫多罗基金会展览馆 |

## 02 Commercial & Entertainment 商业+娱乐休闲

| | | | | |
|---|---|---|---|---|
| 110 | Saffire Tasmania<br>照明与生态的融合——塔斯马尼亚岛莎菲尔度假村 | | 168 | RED Prime Steak<br>融入氖灯的红酒文化——别克大厦红色牛排餐厅 |
| 132 | The Island Club<br>岛屿的夜间表情——三亚东锣岛 | | 176 | The Garden Restaurant "CAD" in Kiev<br>隐藏式LED打造的魔幻色彩——基辅花园餐厅 |
| 144 | Surfers Paradise Foreshore<br>线性半导体特色照明——冲浪者天堂海滩改造 | | 178 | Duke of York Square and Headquarters Building<br>白色照明的多样化运用——伦敦约克公爵广场及其总部大楼 |
| 150 | Surfers Paradise Hilton<br>变色照明的视觉体验——冲浪者天堂希尔顿酒店 | | 184 | Tokyo-Harajukal Ginza H&M Stores, Osaka, Seoul, Singapore<br>商业照明的典范——H&M服饰店 |
| 160 | Show Villa at Sanctuary Falls<br>灯光、景观、建筑三位一体——圣地瀑布度假别墅 | | | |

# 03 Transportation 交通

| | | | |
|---|---|---|---|
| 194 | Bi-Tan Bridge<br>光与桥的对话——碧潭吊桥 | 240 | Swivel Bridge, Bremerhaven<br>间接照明的典范——不来梅哈芬平旋桥 |
| 202 | Prins Claus Bridge, Utrecht, The Netherlands<br>低能耗照明——克劳斯王子桥 | 244 | Epping to Chatswood Tail Link<br>精炼与提纯——艾平－查茨伍德换乘平台 |
| 216 | Telekom Bridge<br>物理"光影"体验——电信步行桥 | 254 | Los Angeles World Airports<br>黑夜只是背景——洛杉矶国际机场 |
| 224 | The Rion Antirion Bridge in Greece<br>神秘而壮观的梦中夜景——希腊里奥·安托里恩大桥 | 260 | Underground<br>多种光源在市政建设中的运用——科恩路地道 |
| 234 | Current$^3$<br>会"呼吸"的互动性光照设计——Current$^3$ | | |

# 04 Cityscape & Public Art 城市景观+公共艺术

| | | | |
|---|---|---|---|
| 270 | Clichy Batignolles Park in Paris 17e<br>超低能耗的特殊照明——巴黎17区克利西巴蒂诺尔公园 | 310 | Slater Mill Falls<br>白色光照的梦幻效果——斯莱特米尔瀑布 |
| 276 | Robert F. Kennedy Inspiration Park<br>背景灯光的视觉盛宴——肯尼迪纪念公园 | 314 | Shed 1, Princess Wharf<br>悬浮的"交流屏"——公主码头1号棚 |
| 282 | Wolfsburg Automobile City<br>光影交错——沃尔夫斯堡汽车城 | 326 | Louisville Second Street<br>落日的色彩——路易斯维尔第二街道交通和街景工程 |
| 292 | An Eventful Path<br>足下的艺术——悉尼奥林匹克公园历史大道 | 332 | Frederiksberg New Urban Spaces<br>铺展的画卷——菲德烈堡新城市空间 |
| 298 | Glostrup Town Hall Park<br>星空下的童话王国——Glostrup市镇厅公园 | 340 | 10th@Hoyt Apartments<br>点亮生活的夜晚——霍伊特公寓 |
| 304 | The Garonne in Toulouse<br>新技术演绎的生态照明——图卢兹市加伦河 | 346 | The Albany Courtyard "Garden of Light"<br>光影欢乐园——阿尔伯尼庭院 |

# 05 Thesis 先锋论点

**412** Night Landscape
灯光与自然的和谐共处——景观照明

**420** LED, Sustainability, Preservation and Art
保护与艺术——LED灯的可持续性

---

**350** "Vessel"
光之篮——西雅图弗莱德·哈钦森癌症研究中心"容器"雕塑

**358** City Lights Streetlight—Fabrication of Prototypes
城市之光——城市街道照明设施雏型制作

**364** "Tecotosh" Portland State University
结构与光线的互动——波特兰州立大学"Tecotosh"雕塑

**372** Lauderdale-Hollywood Airport, Florida
"鳍扇"——劳德代尔堡-好莱坞国际机场雕塑

**376** "Triplet"
悬浮的三角研究园——罗利达勒姆国际机场三联雕塑

**382** Gifu Kitagata Apartments
光、色、艺术、生活——日本岐阜kitagata公寓

**386** The Port Pavilion
光影潮汐——圣地亚哥百老汇港口大厅

**390** "De Beemd" in Velp
光束波浪舞——威尔普德比姆工业区

**392** The Light Orchestra
交互式艺术装置——光乐团

**396** Alto Calore Aqueduct
深蓝里的黑色影像——高山上的"水之萨满"雕塑

**404** East Fremont Street
霓虹游乐场——弗里蒙特东街街区

# Overview 综述

This section will focus on the concept and types of lighting, lighting industry standard, green lighting, and so on.
本节，将围绕着照明的概念与类型、照明的行业标准、绿色照明等方面展开阐述。

## 一、Concept and Types of Lighting
### 照明的概念与种类

"Cihai" defined lighting as：measures of using the light source to illuminate the working and living place, or individual objects. Sun and sky as light source are named"natural lighting"; Man-made light source is named" artificial lighting". The primary purpose of lighting is to create a good visibility and a comfortable and enjoyable environment. There are normal lighting, emergency lighting, duty lighting, security lighting and obstruction lighting. Emergency lighting includes the spare lighting, security lighting and evacuation lighting.

《辞海》将照明定义为：利用光源照亮工作和生活场所或个别物体的措施。利用太阳和天空作为光源的称为"天然照明"；利用人工光源的称为"人工照明"。照明的首要目的是创造良好的可见度和舒适愉快的环境。

照明种类可分为：正常照明、应急照明、值班照明、警卫照明和障碍照明。其中应急照明包括备用照明、安全照明和疏散照明。

## 二、Basic patterns of lighting
### 照明的基本方式

Patterns of lighting: according to installation location or light distribution, the basic standard of the lighting. According to light distribution and lighting effects there are direct lighting and indirect lighting. According to installation location, there are general lighting(include district general lighting), local lighting and mixed lighting.

照明方式指照明设备按其安装部位或光的分布而构成的基本制式。按光的分布和照明效果可分为直接照明和间接照明。就安装部位而言，有一般照明(包括分区一般照明)、局部照明和混合照明等。

### (一) Direct lighting and indirect lighting
#### 直接照明和间接照明

Direct lighting：Top 0～10%、Below100～90% Light distribution，Opaque reflective umbrella，e.g.Spotlight.
直接照明：上方0～10%、下方100～90%的配光，不透明反射伞，例：射灯。
Semi-direct lighting：Top10～14%、Below90～60% Light distribution，Translucent umbrella，e.g.Pendant lamp
半直接照明：上方10～14%、下方90～60%的配光，半透明伞，例：吊灯。
All diffused lighting：Top40～60%、Below60～40% Light distribution，e.g.Dome lamp.
全方位扩散照明：上方40～60%、下方60～40%的配光，例：球型灯。
Semi-indirect lighting：Top60～90%、Below40～10% Light distribution，Translucent reflective umbrella，e.g.Door lights.
半间接照明：上方60～90%、下方40～10%的配光，半透明反射伞，例：门灯。
Indirect lighting：Top90～100%、Below10～0% Light distribution，Opaque reflective umbrella，e.g.Wall lamp.
间接照明：上方90～100%、下方10～0%的配光，不透明反射伞，例：壁灯。

### (二)General lighting and Local lighting
#### 一般照明与局部照明

General lighting：Uniform lighting to illuminate the entire premises.
一般照明：为照亮整个场所的均匀照明。
Local lighting：Lighting for illuminating a specific visual working and a local setting.
局部照明：特定视觉工作用的、为照亮某个局部而设置的照明。
Mixed lighting：Consisting of general lighting and local lighting.
混合照明：由一般照明与局部照明组成的照明。

## 三、Lighting Industry Standards
### 照明的行业标准

### (一)Lighting technology standards of main light source
#### 主要光源的照明技术标准

| Light source types 光源种类 | Luminous efficiency 光效(Lm/W) | Color rendering index 显色指数Ra | Color temperature 色温(K) | Average life 平均寿命(h) |
|---|---|---|---|---|
| Incandescent light bulbs 白炽灯泡 | 15 | 100 | 2 800 | 1 000 |
| Quartz halogen 石英卤素灯 | 25 | 100 | 3 000 | 2 000-3 000 |
| SL light SL灯 | 50 | 87 | 2 700/5 000 | 8 000 |
| High pressure mercury lamp 高压汞灯 | 50 | 45 | 3 300-4 300 | 6 000 |
| Ordinary fluorescent 普通日光灯 | 70 | 70 | Full series 全系列 | 8 000 |
| PL type lamp PL型灯管 | 85 | 85 | 2 700/3 000/3 500/ 4 000/5 000/5 300 | 8 000-12 000 |

To be Continued 续前表

| | | | | |
|---|---|---|---|---|
| Metal halide lamps<br>金属卤化物灯 | 75-95 | 65-92 | 3 000/4 500/5 600 | 6 000-20 000 |
| Pansy lamps<br>三基色日光灯 | 96 | 80-98 | Full series<br>全系列 | 10 000 |
| High pressure sodium lamps<br>高压钠灯 | 120 | 23/60/85 | 1 950/2 200/2 500 | 24 000 |
| Low pressure sodium lamp<br>低压钠灯 | 200 | 44 | 1 700 | 28 000 |
| QL lamp<br>QL灯 | 70 | 85 | 3 000/4 000 | 80 000 |

(二) Lighting color temperature and the human eye kruith curve
灯光色温与人眼舒适度曲线 (Kruithof舒适性曲线)

A lower color temperature lighting system(2,700- 3,000K)between 50 to100 lx illumination can be able to offer comfortable lighting.The illumination by further increasing will cause unpleasantness.
色温较低的照明系统(2 700- 3 000K)在50至100 lx的照度下能提供舒适的照明，而如果照度进一步提高，将会令人感到不快。

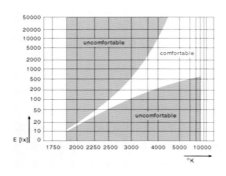

(三) National illumination standards comparison
各国照度标准比较

National illumination standards values are classified according to 0.5、1、3、5、10、15、20、30、50、75、100、150、200、300、500、750、1,000、1,500、2,000、3,000、5,000lx. lx is the unit of illumination. National illumination standards are different from each other; the following below is a detailed comparison.
照度标准值按0.5、1、3、5、10、15、20、30、50、75、100、150、200、300、500、750、1 000、1 500、2 000、3 000、5 000lx 分级， 以lx(勒克斯)为照度单位。各国照度标准不尽相同，以下做一个详细比较。

1. National Civil illumination standards
各国民用建筑照度标准

This chapter discusses the civil illumination standard, including public buildings, refers to office buildings, commercial buildings, tourism buildings, science, education buildings, transportation buildings and residential buildings illumination standards.
本章论述的民用建筑照度标准包括公共建筑即办公建筑、商业建筑、旅游建筑、科教文卫建筑以及交通运输建筑和居住建筑的照度标准。

(1)Illumination standards of office building domestic and abroad
办公建筑国内外照度标准

### Illumination values standard comparison of office building illumination standards (unit: lx)
办公建筑国内外照度标准值对比(单位: lx)

| Room or Site<br>房间或场所 | China中国<br>GB 50034-2004 | CIE<br>S 008/E-2001 | USA美国<br>IESNA-2000 | Japan日本<br>JISZ9100-1979 | Germany德国<br>DIN5035-1990 | sia俄罗斯<br>CHNII23-05-95 |
|---|---|---|---|---|---|---|
| General office<br>普通办公室 | 300<br>500 | 500 | 500 | 300~750 | 300<br>500 | 300 |
| High-grade office<br>高档办公室 | | | | | | |

To be Continued 续前表

| Meeting room Anteroom reception 会议室、接待室、前台 | 300 | 500<br>300<br>(Anteroom 接待) | 300<br>500<br>(Important 重要) | 300~750<br>200~500<br>(Anteroom 接待) | 300 | 200<br>300<br>(Reception 前台) |
|---|---|---|---|---|---|---|
| Business hall 营业厅 | 300 | —— | 300<br>500<br>(Writing 书写) | 750~1 500 | —— | —— |
| Design Room 设计室 | 500 | 750 | 750 | 750~1 500 | 750 | 500 |
| Files, copy, mail room 文件整理、复印、发行室 | 300 | 300 | 100 | 300~750 | —— | 400 |
| Material, document room 资料、档案室 | 200 | 200 | —— | 150~300 | —— | 75 |

(2) Illumination standards of commercial building domestic and abroad
商业建筑国内外照度标准

Because of different service objects, goods grade, and decoration requirements, illumination is also varied, each country has its own features.
鉴于各类商店服务对象的不同，商品档次不同，装饰要求不同，对照度的要求也不同，各国均有自己的特点。

Illumination values standard comparison of the commercial buildings domestic and abroad (unit: lx)
商业建筑国内外照度标准值对比(单位: lx)

| Room or site 房间或场所 | China中国 GB 50034-2004 | CIE S 008/E-2001 | USA美国 IESNA-2000 | Japan日本 JISZ9100-1979 | Germany德国 DIN5035-1990 | Russia俄罗斯 CHNII23-05-95 |
|---|---|---|---|---|---|---|
| General business hall 一般商业营业厅 | 300 | 300(Small 小) | 300 | 500~750 | 300 | 300 |
| High-end business hall 高档商业营业厅 | 500 | 500(Big 大) | | | | |
| General supermarket hall 一般超市营业厅 | 300 | —— | 500 | 750~1 000<br>(the city 市内)<br>300~750<br>(the outskirts 郊外) | —— | 400 |
| High-end supermarket hall 高档超市营业厅 | 500 | | | | | |
| Cashier 收款台 | 500 | 500 | —— | 750~1 000 | 500 | —— |

(3) Illumination standards of hotel building domestic and abroad
旅馆建筑国内外照度标准

Illumination values standard comparison of hotel building illumination domestic and abroad (unit: lx)
旅馆建筑国内外照度标准值对比(单位: lx)

| Room or site 房间或场所 | China中国 GB 50034-2004 | CIE S 008/E-2001 | USA美国 IESNA-2000 | Japan日本 JISZ9100-1979 | Germany德国 DIN5035-199 | Russia俄罗斯 CHNII23-05-95 |
|---|---|---|---|---|---|---|

To be Continued 续前表

| Room | Sub-area | China | CIE | USA | Japan | Germany | Russia |
|---|---|---|---|---|---|---|---|
| Guest house 客房 | Generally activity area 一般活动区 | 75 | | 100 | 100~150 | | 100 |
| | Bedside 床头 | 150 | — | — | — | — | — |
| | Desk 写字台 | 300 | — | 300 | 300~750 | — | — |
| | Bathroom 卫生间 | 150 | — | 300 | 100~200 | — | — |
| Chinese Restaurant 中餐厅 | | 200 | 200 | — | 200~300 | 200 | — |
| Western restaurants, bars, cafes 西餐厅、酒吧间、咖啡厅 | | 100 | — | — | — | — | — |
| Multi-function hall 多功能厅 | | 300 | 200 | 500 | 200~500 | 200 | 200 |
| Hall, the service desk and Lounge 门厅、总服务台、休息厅 | | 300 / 200 | 300 | 100 / 300 (Reading 阅读处) | 100~200 | — | — |
| Rooms layer Corridor 客房层走廊 | | 50 | 100 | 50 | 75~100 | — | — |
| Kitchen 厨房 | | 200 | — | 200~500 | — | 500 | 200 |
| Laundry 洗衣房 | | 200 | — | — | 100~200 | — | 200 |

(4) Illumination standards of school building domestic and abroad
学校建筑国内外照度标准

Illumination values standard comparison of school building domestic and abroad (unit: lx)
学校建筑国内外照度标准值对比(单位: lx)

| Room or site 房间或场所 | China 中国 GB 50034-2004 | CIE S 008/E-2001 | USA 美国 IESNA-2001 | Japan 日本 JISZ9110-1979 | Germany 德国 DIN5035-1990 | Russia 俄罗斯 CHNII 23-05-95 |
|---|---|---|---|---|---|---|
| Classroom 教室 | 300 | 300 500 (Evening schools, adult education 夜校、成人教育) | 500 | 200~750 | 300~500 | 300 |
| Laboratory 实验室 | 300 | 500 | 500 | 200~750 | 500 | 300 |
| The art room 美术教室 | 500 | 500~750 | 500 | — | 500 | — |

To be Continued 续前表

| | | | | | |
|---|---|---|---|---|---|
| Multimedia classrooms 多媒体教室 | 300 | 500 | —— | —— | 500 | 400 |
| Classroom blackboard 教室黑板 | 500 | 500 | —— | —— | —— | 500 |

(5) Illumination standards of library building domestic and abroad
图书馆建筑国内外照度标准

Illumination values standard comparison of library building domestic and abroad (unit: lx)
图书馆建筑国内外照度标准值对比(单位: lx)

| Room or site 房间或场所 | | China中国 GB 50034-2004 | CIE S 008/E-2001 | USA美国 IESNA-2001 | Russia俄罗斯 CHNII23-05-95 |
|---|---|---|---|---|---|
| Reading room 阅览室 | General library 一般图书馆 | 300 | 500 | 300 | 300 (General一般) |
| | National, provincial and other important libraries 国家、省市及其它重要图书馆 | 500 | | | |
| Older reading, rare books, Maps reading room 老年阅览室、珍善本、舆图阅览室 | | 500 | —— | 300 | —— |
| Directory Hall (room), showroom 目录厅(室)、陈列室 | | 300 | 200 (Personal Bookshelf 个人书架) | 300 (Reading shelf 阅读架) | 200 |
| Stacks 书库 | | 50 | 200 (Bookshelf 书架) | 50 (Fixed 固定设备) | 75 |
| Workplace 工作间 | | 300 | —— | —— | 200 |

(6) Illumination standards of museum building showroom exhibits domestic and abroad
博物馆建筑陈列室展品国内外照度标准

Illumination value standard comparison of museum building showroom exhibits domestic and abroad (unit: lx)
博物馆建筑陈列室展品国内外照度标准值对比(单位: lx)

| Category类别 | China中国 GB 50034-2004 | USA美国 IESNA-2000 | Japan日本 JISZ9110-1979 | UK英国 GIBS-1984 | Russia俄罗斯 CHNII 23-05-95 |
|---|---|---|---|---|---|
| Exhibits particularly sensitive to light 对光特别敏感的展品 | 50 | —— | 75~150 | 50 | 50~75 |
| Exhibits sensitive to light 对光敏感的展品 | 150 | —— | 300~750 | 150 | 150 |
| Exhibits not sensitive to light 对光不敏感的展品 | 300 | —— | 750~1 500 | —— | 200~500 |

(7) Illumination standards of exhibition hall domestic and abroad
展览馆展厅国内外照度标准

Illumination standards value comparison of exhibition hall domestic and abroad (unit: lx)
展览馆展厅国内外照度标准值对比(单位: lx)

| Category 类别 | | China 中国 GB 50034-2004 | USA 美国 IESNA-2000 | Japan 日本 JISZ9110-1979 | Russia 俄罗斯 CHNII 23-05-95 |
|---|---|---|---|---|---|
| Exhibition Hall 展厅 | General 一般 | 200 | 100 | 200~500 | 200 |
| | High-end 高档 | 300 | | | |

(8) Illumination standards of theater and cinema building domestic and abroad
影剧院建筑国内外照度标准

Illumination value standard comparison of theater and cinema building domestic and abroad (unit: lx)
影剧院建筑国内外照度标准值对比(单位: lx)

| Room or Site 房间或场所 | | China 中国 GB 50034-2004 | CIE S 008/E-2001 | USA 美国 IESNA-2000 | Japan 日本 JIS Z9100-1979 | Russia 俄罗斯 CHNII23-05-95 |
|---|---|---|---|---|---|---|
| the Entrance hall 门厅 | | 200 | 100 | —— | 300~750 | 500 |
| Audience Hall 观众厅 | Cinema 影院 | 100 | 200 | 100 | 150~300 | 75 |
| | Theater 剧院 | 200 | | | | 300~500 |
| Audience' Lounge 观众休息厅 | Cinema 影院 | 150 | —— | —— | 150~300 | 150 |
| | Theater 剧院 | 200 | | | | —— |
| Rehearsal Hall 排演厅 | | 300 | 300 | —— | —— | —— |
| the Dressing room 化妆室 | General activities area 一般活动区 | 150 | —— | —— | 300~750 | —— |
| | dresser 化妆台 | 500 | | | | |

(9) Illumination standards of sports building domestic and abroad
体育建筑照度国内外照度标准

Lighting is a very important aspect in the construction of the modern multi-functional stadium, it meets not only the requirements of various sports competitions on the site of illumination, but also controls light and color, glare to reach a certain international standards. Auditorium and other area lighting are required to adapt to different needs of various environments, achieve different lighting effects.

在现代化多功能体育场馆的建设中，照明是十分重要的一个环节，它不仅要满足各类体育比赛对场地的照度要求，而且对照度、光色、眩光都要达到一定的国际标准。对于观众席及其他区域照明则要求能适应不同环境的不同需求，达到不同的照明效果。

Illumination value standards comparison of sports building domestic and abroad (unit: lx)
体育建筑照度国内外照度标准值对比(单位: lx)

| Category 类别 | CIE No 83-1989 | USA 美国 IESNA-2000 | Japan 日本 JISZ 9110-1979 |
|---|---|---|---|
| Stadium 体育场 | 500~750~100(A) 750~100~140(B) 1 000~140(C) | 1 000~1 500 | 750~1 500 (Formal 正式) 300~750 (General 一般) |
| Gym 体育馆 | | 1 500~2 000 | 750~1 500 (Formal 正式) 300~750 (General 一般) |
| Swimming pool 游泳馆 | | 300~750 | 750~1 500 (Formal 正式) 300~750 (General 一般) |

Note: CIE standards (A), (B) and (C) are TV broadcasting of illumination values of three group competition games.
注: CIE标准的(A)、(B)和(C)为三组比赛项目的彩电转播照度值。

Each country sports building illumination standards are varied, China has more detailed regulations, which provides a variety of sports illumination value standard, and develops different illumination of the TV broadcast and Non-TV broadcast standards, as follows:

各国对于体育建筑的照度标准各不相同，中国则做了更为详尽的规定，规定了各种运动项目所对应的照度标准值，并制定了无彩电转播和有彩电转播的不同照度标准，具体情况如下：

## Illumination value standards of sports building without TV broadcast (unit: lx)
### 无彩电转播的体育建筑照度标准值 (中国GB 50034-2004)

| Sports 运动项目 | Reference surface 参考平面 | Illumination value Standards 照度值标准(Lx) | |
| --- | --- | --- | --- |
| | | Training 训练 | Competition 比赛 |
| Basketball, Volleyball, Badminton, Tennis, Handball, track and field (indoor) gymnastics, rhythmic gymnastics, techniques, martial 篮球、排球、羽毛球、网球、手球、田径和(室内)体操、艺术体操、技巧、武术 | Ground 地面 | 300 | 750 |
| Baseball, softball 棒球、垒球 | Ground 地面 | —— | 750 |
| Bowling 保龄球 | Bottle area 瓶置区 | 300 | 500 |
| Weight lifting 举重 | Stage 台面 | 200 | 750 |
| Fencing 击剑 | Stage 台面 | 500 | 750 |
| Judo, Chinese wrestling and international wrestling 柔道、中国摔跤、国际摔跤 | Ground 地面 | 500 | 1 000 |
| Boxing 拳击 | Stage 台面 | 500 | 2 000 |
| Table tennis 乒乓球 | Stage 台面 | 750 | 1 000 |
| Swimming, fin swimming, diving, water polo 游泳、蹼泳、跳水、水球 | Water surface 水面 | 300 | 750 |
| Synchronized swimming 花样游泳 | Water surface 水面 | 500 | 750 |
| Ice hockey, speed skating, figure skating 冰球、速度滑冰、花样滑冰 | Ice surface 冰面 | 300 | 1 500 |
| The game of go, Chinese chess, international chess 围棋、中国象棋、国际象棋 | Stage 台面 | 300 | 750 |
| Bridge 桥牌 | Stage 桌面 | 300 | 500 |
| Shooting 射击 | Target vertical surface 靶心垂直面 | 1 000 | 1 500 |
| | Ground 地面 | 300 | 500 |
| Football, hockey 足球、曲棍球 | Ground 地面 | —— | 300~750 |
| the Auditorium 观众席 | Seating surface 座位面 | —— | 100 |

To be Continued 续前表

| | | | |
|---|---|---|---|
| Gym 健身房 | Ground 地面 | 200 | —— |

**Illumination value standards of sports building with TV broadcast (unit: lx)**
**有彩电转播的体育建筑照度标准值 (中国GB 50034-2004)**

| Sports 运动项目 | Reference surface and height 参考平面及其高度 | Illumination value standards(lx) 照度标准值(lx) Maximum PD(m) 最大摄影距离(m) | | |
|---|---|---|---|---|
| | | 25 | 75 | 150 |
| A: Wrestling, judo, swimming, track and field A组：田径、柔道、游泳、摔跤等项目 | 1.0m vertical surface 1.0米垂直面 | 500 | 750 | 1 000 |
| B: Basketball, tennis, handball, volleyball, badminton, gymnastics, figure skating, speed skating, baseball, football... B组：篮球、排球、羽毛球、网球、手球、体操、花样滑冰、速滑、垒球、足球等项目 | 1.0m vertical surface 1.0米垂直面 | 750 | 1 000 | 1 500 |
| C：Boxing, fencing, diving, table tennis, ice hockey... C组：拳击、击剑、跳水、乒乓球、冰球等项目 | 1.0m vertical surface 1.0米垂直面 | 1 000 | 1 500 | —— |

(10) Illumination standards of hospital building domestic and abroad
医院建筑国内外照度标准

Light source and the surrounding environment color are essential hospital lighting issue. Hospital lighting is to give full play to the function of hospital work, either directly or indirectly. Hospital where the workplace holds many of the activities, so the illumination varies, such as precision operation requires a higher level of illumination in the operating room, general illumination of the ward is at 100-300lx. Recommended illumination show of all hospital departments are as follows:

光源和周围环境的色彩是医院照明中的重要问题。医院照明是以能充分发挥医院的功能，直接或间接地对医院起作用为目的。医院的工作场所的活动地方很多，照度的差别很大。如精密作业的手术室需要较高的照度，一般病房的照度在100-300lx即可。现把各国医院各部门的推荐照度展示如下：

**Illumination value standards comparison of hospital architectural domestic and abroad (unit: lx)**
**医院建筑国内外照度标准值对比(单位: lx)**

| Room or site 房间或场所 | China industry standard 中国 行业标准 JGJ 49-88 | China industry standard 2004 2004 试行版 | China中国 GB 50034-2004 | CIE S 008/E-2001 | USA美国 IESNA-2000 | Japan日本 JIS Z9100-1979 | Germany德国 DIN5035-1990 |
|---|---|---|---|---|---|---|---|
| Treatment room 治疗室 | 50~100 | 150 | 300 | 1 000 500 (General一般) | 300 | 300~750 | 300 |
| Laboratory 化验室 | 75~150 | 200 | 500 | 500 | 500 | 200~500 | 500 |
| Operating room 手术室 | 100~200 | 500 | 750 | 1 000 | 3 000~10 000 | 750~1 500 | 1 000 |
| Consulting room 诊室 | 75~150 | 150~200 | 300 | 500 | 300(General一般) 500(Working table工作台) | 300~750 | 500~1 000 |
| Waiting room, registered office 候诊室、挂号厅 | 50~100 | 150 | 200 | 200 | 100(General一般) 300(Reading阅读) | 150~300 | —— |
| Ward 病房 | 15~30 | 50 | 100 | 100(General一般) 300(Examination检查) 300(Reading阅读) | 50(General一般) 300(Reading阅读) 500(Diagnosis诊断) | 100~200 | 100(General一般) 200(Reading阅读) 300(Examination检查) |

To be Continued 续前表

| Room or site 房间或场所 | | | | | |
|---|---|---|---|---|---|
| Nurse station 护士站 | | 75~150 | 200 | 300 | — | 300(General一般) 500(Table桌面) | 300~750 | 300 |
| Pharmacy 药房 | | — | — | 500 | — | 500 | 300~750 | — |
| Intensive care room 重症监护室 | | — | — | 300 | 500 | — | — | 300 |

(11) Illumination standard of transportation building domestic and abroad
　　交通建筑国内外照度标准

Transportation plays an important role in the modern society economic development and people's lives, in accordance with different modes of transport, there are four kinds including train station (rail), bus station (road), airport (aviation) and terminal (water transport). In the passenger transport, ticket selling, waiting areas, ticket checking, baggage and other aspects, passengers' satisfaction on visual aspects can be achieved through a suitable lighting system.

交通运输在现代社会经济发展和人民的生活中起着重要作用，按照运输方式不同，大致可以分为火车站（铁路）、汽车站（公路）、机场（航空）以及码头（水运）四种。在交通客运中，售票、候车、检票、行李托运等各个环节都需要通过合理的照明系统来满足乘客在视觉方面的各种不同需求。

Illumination value standards comparison of traffic building(locomotive, bus station, airport, terminal)domestic and abroad (unit: lx)
交通建筑(火车头、汽车站、机场、码头)国内外照度标准值对比 (单位：lx)

| Room or site 房间或场所 | | China中国 GB 50034-2004 | CIE S 008/E -2001 | USA美国 IESNA-2000 | Japan日本 JISZ9110-1979 |
|---|---|---|---|---|---|
| Ticket counter 售票台 | | 500 | — | — | — |
| The information desk 问讯处 | | 200 | 500(Counter台面) | — | — |
| Waiting room (machines, ship) 候车(机、船)室 | General 普通 | 150 | 200 | 50 | 300~175(A) 150~300(B) 75~150(C) |
| | High-end 高档 | 200 | | | |
| Central Hall 中央大厅 | | 200 | 200 | 30 | — |
| the booking Hall 售票大厅 | | 200 | 200 | 500 | 300~750(A) 150~300(B) |
| Customs and passport control 海关、护照检查 | | 500 | 500 | — | — |
| Security check 安全检查 | | 300 | 300 | 300 | — |
| Tickets exchange, baggage 换票、行李托运 | | 300 | 300 | 300 | — |
| Baggage reclaim, arrival hall, departure hall 行李认领、到达大厅、出发大厅 | | 200 | 200 | 50 | — |
| Channel, link, ladder 通道、连接区、扶梯 | | 150 | 150 | — | 150~300(A) 75~150(B) 50~150(C) |
| Platform (no canopy) 站台(有篷) | | 75 | — | — | 150~300(A) |
| Platform (canopy) 站台(无篷) | | 50 | — | — | 75~150(B) |

(12) Illumination standard of residential buildings domestic and abroad
居住建筑国内外照度标准

Residential buildings design should take full advantage of the sunshine conditions provided by the external environment, inner space should be able to provide the compatible illumination, illumination in bedrooms, kitchens, and other places, should first meet its requirements of usage function.

居住建筑在设计时，都应充分利用外部环境提供的日照条件，套内的空间应能提供与其使用功能相适应的照度水平，卧室、厨房等的照度都应首先满足其使用功能需求。

Illumination value standards comparison of residential buildings domestic and abroad (unit: lx)
居住建筑国内外照度标准值对比（单位：lx）

| Room or site 房间或场所 | | China 中国 GB 50034-2004 | Japan 日本 JISZ9110-1979 | USA 美国 IESNA-2000 | Russia 俄罗斯 CHNII23-05-95 |
|---|---|---|---|---|---|
| The living room 起居室 | General activities 一般活动 | 100 | 30~75 (General 一般) 150~300 (Focus 重点) | 300 (Read occasionally 偶尔阅读) 500 (Read carefully 认真阅读) | 100 |
| | Writing, reading 书写、阅读 | 300 | | | |
| Bedroom 卧室 | General activities 一般活动 | 75 | 1~30 (General 一般) 300~750 (Reading 读书) 300~750 (make-up 化妆) | 300 (Read occasionally 偶尔阅读) 500 (Read carefully 认真阅读) | 100 |
| | Bedside, reading 床头、阅读 | 150 | | | |
| Dining 餐厅 | | 150 | 50~100 (General 一般) 200~500 (Dining table 餐桌) | 50 | —— |
| Kitchen 厨房 | General 一般活动 | 100 | 50~100 (General 一般) 200~500 (Cook, sink 烹调、水槽) | 300 (General 一般) 500 (Focus 重点) | 100 |
| | Operation panel 操作台 | 150 | | | |
| Bathroom 卫生间 | | 100 | 75~150 (General 一般) 200~500 (face washing, make-up 洗脸、化妆) | 300 | 50 |

2、Illumination standard of Industrial building domestic and abroad
工业建筑国内外照度标准

Industrial lighting design principles are safety, utility, economy and beauty. As green lighting, as the promotion of green, energy-saving, more and more industries, enterprises pay attention to their own corporate image, for better lighting design of industrial establishments puts forward higher requirements.

工业照明设计的原则是安全、适用、经济及美观。随着绿色照明、节能照明的推广以及越来越多的工业、企业重视自己的企业形象，如何做好工业场所的照明设计对设计人员提出了更高的要求。

(1) Illumination standard of general rooms or sites domestic and abroad
通用房间或场所国内外照度标准

Illumination value standards comparison of general rooms or sites domestic and abroad (unit: lx)
通用房间或场所国内外照度标准值对比（单位：lx）

| Room or site 房间或场所 | | China 中国 GB 50034-2004 | CIE s 008/E-2001 | Germany 德国 DIN 5035-1990 | USA 美国 IESNA-2000 | Japan 日本 JISZ9110-1979 | Russia 俄罗斯 CHNII23-05-95 |
|---|---|---|---|---|---|---|---|
| Laboratory 实验室 | General 一般 | 300 | 500 | 300 | —— | 300 | —— |
| | Fine 精细 | 500 | —— | | | 3000 | |

To be Continued 续前表

| | | | | | | | |
|---|---|---|---|---|---|---|---|
| Test 检验 | General 一般 | 300 | 750~1000 | 750 | 300~1 000 | 300~31 000 | 200 |
| | Fine with color requirements 精细，有颜色要求 | 750 | | ——— | 3000~101 000 | ——— | ——— |
| Weight room, measuring room 计量室，测量室 | | 500 | 500 | ——— | ——— | ——— | ——— |
| Distribution station 变配电站 | Power distribution room 配电装置室 | 200 | 200~500 | 100 | 500,300,100 | 100,300 | 150,200 |
| | Transformer room 变压器室 | 100 | | | | | 75 |
| Power supply, Generator room 电源设备室，发电机室 | | 200 | 200 | 100 | 500,300,100 | 150,300 | 150,200 |
| Control room 控制室 | General control room 一般控制室 | 300 | 300 | ——— | 100 | 300 | 150,300 |
| | Main control room 主控制室 | 500 | 500 | ——— | ——— | 750 | ——— |
| Telephone stations, Network Center 电话站、网络中心 | | 500 | ——— | 300 | 500,300,100 | ——— | 150,200 |
| Computer station 计算机站 | | 500 | 500 | ——— | 500,300,100 | ——— | ——— |
| The power station 动力站 | Wind the engine room, air conditioning room 风机房、空调机房 | 100 | 200 | 100 | 500,300,100 | 150-300 | 50 |
| | Pump room 泵房 | 100 | | | | | 150,200 |
| | Refrigerating station 冷冻站 | 150 | | | | | ——— |
| | Compressed air station 压缩空气站 | 150 | ——— | ——— | | | 150,200 |
| | Leaking boiler room operation, gas station 漏炉房、煤气站 的操作层 | 100 | 100 | 100 | | | 50~150 |
| Warehouse 仓库 | Large libraries (such as billet, rolled steel, finished products, gas cylinders) 大件库(如钢坯、钢材、大成品、气瓶) | 50 | 100 | 50 | 50 | 30 | 50 |
| | General library 一般件库 | 100 | | 100 | 100 | 50 | 75 |
| | Fine library: tools, small parts 精细件库:如工具、小零件 | 200 | | 200 | 300 | 75 | 200 |
| Vehicle service station 车辆加油站 | | 100 | ——— | 100 | ——— | ——— | ——— |

(2) Illumination standard of mechanical and electrical industry domestic and abroad
机电工业国内外照度标准

**Illumination value standards comparison of mechanical and electrical industry domestic and abroad (unit: lx)**
机电工业国内外照度标准值对比（单位：lx）

| Room or site 房间或场所 | | China中国 GB 50034-2004 | CIE S 008/E-2001 | Germany德国 DIN 5035-1990 | USA美国 IESNA-2000 | Japan日本 JISZ9110-1979 | Russia俄罗斯 CHNII23-05-95 |
|---|---|---|---|---|---|---|---|
| Mechanical processing 机械加工 | Roughing 粗加工 | 200 | —— | —— | 300 | 300 | 200(1 000) |
| | General machining tolerance ≥0.1cm 一般加工公差≥0.1cm | 300 | 300 | 300 | 500 | 750 | 200(1 500) |
| | Precision machining tolerance <0.1cm 精密加工公差<0.1cm | 500 | 500 | 500 | 3 000~10 000 | 1 500~3 000 | 200(2 000) |
| Mechanical and electrical, instrument Assembly 机电、仪表装配 | Large 大件 | 200 | 200 | 200 | 300 | 300 | 200(500) |
| | General parts 一般件 | 300 | 300 | 300 | 500 | —— | 300(750) |
| | Precision 精密 | 500 | 500 | 500 | 3 000~10 000 | 3 000 | —— |
| | Super precision 特精密 | 750 | | | | | |
| Wire and cable manufacturing 电线、电缆制造 | | 300 | 300 | 300 | —— | —— | —— |
| Coil winding 线圈绕制 | Large coils 大线圈 | 300 | 300 | 300 | —— | —— | —— |
| | Secondary coils 中等线圈 | 500 | 500 | 500 | —— | —— | —— |
| | Fine coil 精细线圈 | 750 | 750 | 1000 | —— | —— | —— |
| Coil casting 线圈浇注 | | 300 | 300 | 300 | —— | —— | —— |
| Welding 焊接 | General 一般 | 200 | 300 | 300 | 300 | 200 | 200 |
| | Precision 精密 | 300 | 300 | 300 | 3 000~10 000 | 200 | 200 |
| Sheet metal 板金 | | 300 | 300 | 300 | —— | —— | —— |
| Stamping, cut 冲压、剪切 | | 300 | 300 | 200 | 300,500,1 000 | —— | —— |
| Heat treatment 热处理 | | 200 | —— | —— | —— | —— | —— |
| Casting 铸造 | Melting, casting 熔化、浇铸 | 200 | 300,200 | 300,200 | —— | —— | —— |
| | Modeling 造型 | 300 | 500 | 500 | —— | —— | —— |
| Precision casting mould and shelling 精密铸造的制模、脱壳 | | 500 | —— | —— | —— | —— | —— |

| | | | | | | | |
|---|---|---|---|---|---|---|---|
| Drop forging 锻工 | | | 200 | 300,200 | 200 | — | — | 200 |
| Electroplating 电镀 | | | 300 | 300 | 300 | — | — | 200(500) |
| Spray paint 喷漆 | General 一般 | | 300 | 750 | 500 | 300,500,1 000 | — | 200 |
| | Precision 精细 | | 500 | | | | | 300 |
| Pickling, etching, cleaning 酸洗、腐蚀、清洗 | | | 300 | — | — | — | — | — |
| Polishing 抛光 | General 一般装饰性 | | 300 | — | 500 | 300,500,1 000 | — | — |
| | Precision 精细 | | 500 | | | | | |
| Composite materials processing, stack paving, decorative 复合材料加工、铺叠、装饰 | | | 500 | — | — | — | — | — |
| Mechanical and electrical repair 机电修理 | General 一般 | | 200 | — | 200 | 500 | — | 300(750) |
| | Precision 精密 | | 300 | | 500 | | | |

(3) Illumination standard of electronics, textile and chemical fiber industry domestic and abroad
电子、纺织化纤国内外照度标准

Illumination value standards comparison of electronics, textile and chemical fiber industry domestic and abroad (unit: lx)
电子、电子化纤国内外照度标准值对比（单位：lx）

| Room or site 房间或场所 | China中国 GB 50034-2004 | CIE S 008/E-2001 | Germany德国 DIN 5035-1990 | USA美国 IESNA-2000 | Japan日本 JISZ9110-1979 | Russia俄罗斯 CHNII23-05-95 |
|---|---|---|---|---|---|---|
| Electronics industry 电子工业 | | | | | | |
| Electronic components 电子元器件 | 500 | 1 500 | 1 000 | — | 1 500~3 000 | — |
| Electronic parts 电子零部件 | 500 | 1 500 | 1 000 | — | — | — |
| Electronic materials 电子材料 | 300 | — | — | — | — | — |
| Acid, alkali, liquid and powder preparation 酸、碱、药液及粉配制 | 300 | — | — | — | — | — |
| Textile and chemical fiber 纺织化纤 | | | | | | |
| Textile 纺织 | 150~300 | 200~1 000 | 200~1 000 | — | — | — |
| Chemical fiber 化纤 | 75~200 | | | — | — | — |

(4) Illumination standard of pharmaceutical, rubber industry domestic and abroad
制药、橡胶工业国内外照度标准

Illumination value standards comparison of pharmaceutical, rubber industry domestic and abroad (unit: lx)
制药、橡胶工业国内外照度标准值对比 (单位：lx)

| Room or site 房间或场所 | China中国 GB 50034-2004 | CIE S 008/E-2001 | Germany德国 DIN 5035-1990 | USA美国 IESNA-2000 | Japan日本 JISZ9110-1979 | Russia俄罗斯 CHNII23-05-95 |
|---|---|---|---|---|---|---|
| The pharmaceutical industry 制药工业 | | | | | | |
| Pharmaceutical production 制药生产 | 300 | 500 | —— | —— | —— | —— |
| Production flow channel 生产流转通道 | 200 | | —— | —— | —— | —— |
| Rubber industry 橡胶工业 | | | | | | |
| Rubber plant 炼胶车间 | 300 | 500 | —— | —— | —— | —— |
| Rolling and extrusion section 压延、压出工段 | 300 | | —— | —— | —— | —— |
| Forming, cutting section 成型、裁断工段 | 300 | | —— | —— | —— | —— |
| Section of sulfide 硫化工段 | 300 | | —— | —— | —— | —— |

(5) Illumination standard of electric power, iron and steel, paper industry domestic and abroad
电力、钢铁、造纸工业国内外照度标准

Illumination value standards comparison of electric power, iron and steel, paper industry domestic and abroad (unit: lx)
电力、钢铁、造纸工业国内外照度标准值对比 (单位：lx)

| Room or site 房间或场所 | China中国 GB 50034-2004 | CIE S 008/E-2001 | Germany德国 DIN 5035-1990 | USA美国 IESNA-2000 | Japan日本 JISZ9110-1979 | Russia俄罗斯 CHNII23-05-95 |
|---|---|---|---|---|---|---|
| Electric power industry 电力工业 | | | | | | |
| Boiler room 锅炉房 | 100 | 100 | —— | —— | 75 | 100 |
| Generator room 发电机房 | 200 | 200 | 100 | —— | —— | —— |
| Main control room 主控制室 | 500 | 500 | 300 | —— | —— | 150~300 |
| Iron-making 钢铁工业 | | | | | | |
| Steel-making 炼铁 | 30~100 | 200 | 50~200 | —— | —— | —— |
| steelmaking 炼钢 | 150 | 50~200 | 50~200 | —— | —— | —— |
| Continuous casting 连铸 | 150~200 | 50~200 | 50~200 | —— | —— | —— |
| Steel rolling 轧钢 | 50~200 | 300 | 50~200 | —— | —— | —— |

To be Continued 续前表

| Paper industry 造纸工业 ||||||
|---|---|---|---|---|---|---|
| Pulp and paper industry 制浆造纸工业 | 150~500 | 200~500 | 200~500 | —— | —— | —— |

(6)Illumination standard of food and beverage, glass industry domestic and abroad
食品饮料、玻璃工业国内外照度标准

Illumination value standards comparison of food and beverage, glass industry domestic and abroad (unit: lx)
食品饮料、玻璃工业国内外照度标准值对比（单位：lx）

| Room or site 房间或场所 || China中国 GB 50034-2004 | CIE S 008/E-2001 | Germany德国 DIN 5035-1990 | USA美国 IESNA-2000 | Japan日本 JISZ9110-1979 | Russia俄罗斯 CHNII23-05-95 |
|---|---|---|---|---|---|---|---|
| Food and beverage industry 食品饮料工业 ||||||||
| Food食品 | Bakery, candy 糕点、糖果 | 200 | 200~300 | —— | —— | —— | —— |
| | Dairy products Meat products 乳制品、肉制品 | 300 | —— | 200~500 | —— | —— | —— |
| Beverage 饮料 || 300 | —— | —— | —— | —— | —— |
| Beer 啤酒 | Saccharification 糖化 | 200 | 200 | 200 | —— | —— | —— |
| | Fermentation 发酵 | 150 | 200 | 200 | —— | —— | —— |
| | Package 包装 | 150 | 200 | 200 | —— | —— | —— |
| Glass industry 玻璃工业 ||||||||
| Karst system, preparation and annealing 溶制、备料、退火 || 150 | 300 | 300 | —— | —— | —— |
| Furnace 窑炉 || 100 | 50 | 200 | —— | —— | —— |

(7)Illumination standard of cement, leather industry domestic and abroad
水泥、皮革工业国内外照度标准

Illumination value standards comparison of cement, leather industry domestic and abroad (unit: lx)
水泥工业、皮革工业国内外照度标准值对比（单位：lx）

| Room or site 房间或场所 | China中国 GB 50034-2004 | CIE S 008/E-2001 | Germany德国 DIN 5035-1990 | USA美国 IESNA-2000 | Japan日本 JISZ9110-1979 | Russia俄罗斯 CHNII23-05-95 |
|---|---|---|---|---|---|---|
| Cement industry 水泥工业 |||||||
| Main workshops (crushing, grinding, burning, cement grinding of raw materials and packaging) 主要生产车间（破碎、原料粉磨、烧成、水泥粉磨、包装） | 100 | 200~300 | 200 | —— | —— | —— |

To be Continued 续前表

| | | | | | | |
|---|---|---|---|---|---|---|
| Store 储存 | 75 | —— | —— | —— | —— | —— |
| Transport corridors 输送走廊 | 30 | —— | —— | —— | —— | —— |
| Thick billet molding 粗坯成型 | 300 | 300 | 200 | —— | —— | —— |
| Leather industry 皮革工业 | | | | | | |
| Original skin, bath 原皮、水浴 | 200 | 200 | 200 | —— | —— | —— |
| finished 转毂、整理、成品 | 200 | 300 | 300 | —— | —— | —— |
| Drying 干燥 | 100 | —— | —— | —— | —— | —— |

(8) Illumination standard of tobacco, chemical, oil industry domestic and abroad
卷烟、化学、石油工业国内外照度标准

Illumination value standards comparison of tobacco, chemical, oil industry domestic and abroad (unit: lx)
卷烟、化学、石油工业国内外照度标准值对比（单位：lx）

| Room or site 房间或场所 | China中国 GB 50034-2004 | CIE S 008/E-2001 | Germany德国 DIN 5035-1990 | USA美国 IESNA-2000 | Japan日本 JISZ9110-1979 | Russia俄罗斯 CHNII23-05-95 |
|---|---|---|---|---|---|---|
| Tobacco industry 卷烟工业 | | | | | | |
| Silk reeling workshop 制丝车间 | | 200~300 | 200~300 | —— | —— | —— |
| Cigarette、Filter、Package 卷烟、接过滤嘴、包装 | | 500 | 500 | —— | —— | —— |
| Chemical and petroleum industry 化学、石油工业 | | | | | | |
| Production sites 生产场所 | 100 | 50~300 | 50~200 | —— | —— | —— |
| Places of production assistant 生产辅助场所 | 30-75 | | | —— | —— | —— |

(9) Illumination standard of wood and furniture manufacturing industry domestic and abroad
木业和家具制造工业国内外照度标准

Illumination value standards comparison of wood and furniture manufacturing industry domestic and abroad (unit: lx)
木业和家具制造工业国内外照度标准值对比（单位：lx）

| Room or site 房间或场所 | China中国 GB 50034-2004 | CIE S 008/E-2001 | Germany德国 DIN 5035-1990 | USA美国 IESNA-2000 | Japan日本 JISZ9110-1979 | Russia俄罗斯 CHNII23-05-95 |
|---|---|---|---|---|---|---|
| General machining 一般机器加工 | 200 | —— | 300 | 300 | —— | 200(1 000) |
| Fine machining 精细机器加工 | 500 | 500 | 500 | 500,1 000 | —— | |
| Sawn timber 锯木区 | 300 | 300 | 200 | —— | —— | |

To be Continued 续前表

| | | | | | | | |
|---|---|---|---|---|---|---|---|
| Model area 模型区 | General 一般 | 300 | 750 | 500 | —— | —— | 200(1 000) |
| | Fine 精细 | 750 | | | —— | —— | |
| Gluing, Assembly 胶合、组装 | | 300 | 300 | 300 | —— | —— | 200(1 000) |
| Polished, shaped fine woodworking 磨光、异形细木工 | | 750 | 750 | —— | —— | —— | |

3. Illumination standard in public areas domestic and abroad
   公共场所国内外照度标准

In this section public areas are public and industrial buildings in public areas, such as stairs, toilets, bathrooms, baths, and so on, illumination requirements is not the same in different places.
本节所说公共场所是指公共建筑和工业建筑的公共场所，如楼梯、厕所、盥洗室、浴室的等，不同场所的照度要求也不相同。

Illumination value standards comparison of public areas domestic and abroad (unit：lx)
公用场所国内外照度标准值对比 (单位：lx)

| Room or site 房间或场所 | | China中国 GB 50034-2004 | CIE S 008/E-2001 | 美国 IESNA-2000 | Japan日本 JISZ9110-1979 | Germany德国 DIN 5035-1990 | Russia俄罗斯 CHNII23-05-95 |
|---|---|---|---|---|---|---|---|
| The entrance hall 门厅 | | 100(Normal普通) 200(High-end高档) | 100 | 100 | 200~500 | twice of the adjacent room illumination 相邻房间照度的2倍 | 30~150 |
| Corridor, mobile area 走廊、流动区域 | | 50(Normal普通) 100(High-end高档) | 100 | 100 | 100~200 | 50 | 20~75 |
| Stair, Platform 楼梯、平台 | | 30(Normal普通) 75(High-end高档) | 150 | 50 | 100~300 | 100 | 10~100 |
| Escalator 自动扶梯 | | 150 | 150 | 50 | 50~750 | 100 | —— |
| Toilet, bathroom, bath 厕所、盥洗室、浴室 | | 75(Normal普通) 150(High-end高档) | 200 | 50 | 100~200 | 100 | 50~75 |
| Elevator lobby 电梯前厅 | | 75(Normal普通) 150(High-end高档) | —— | | 200~500 | | |
| Lounge 休息室 | | 100 | 100 | 100 | 75~150 | 100 | 50~75 |
| Storeroom, warehouse 储藏室、仓库 | | 100 | 100 | 100 | 75~150 | 50-200 | 75 |
| Garage 车库 | Parking 停车间 | 75 | 75 | | | | |
| | Maintenance 检修间 | 200 | | | | | |

4. Standard of city night view lighting design domestic and abroad
   城市夜景照明设计国内外标准

CIE international Commission on illumination lighting defined night view as "Exterior Lighting for the decoration of the night time urban landscape". A good city night view lighting is not simply beautified or copy something. As described in the paper of the International Commission on illumination, "a bright city is popular, charming and reasonable, it attracts visitors' interest." "Regardless of the light is traditional or modern, slow or fast, a constant or a moment, it runs through the cities and buildings, makes everything shining and bling."

CIE国际照明委员会定义夜景照明为"夜间室内城市景观装饰照明"。一个好的城市夜景照明，并不是单一的美化亮化，不是简单的模仿、照搬。正如国际照明委员会文件所述"一个明亮的城市是受欢迎、使人陶醉和合乎人意的。它能吸引游客并能诱发他们的兴趣。""无论光线是传统的还是现代的，是慢的还是快的，是恒定不变的还是一闪而过的，它穿过城市上空和建筑，在他们沐浴下，一切都照耀生辉。"

(1) Standard of city night view lighting recommended by CIE international Commission on illumination lighting
CIE国际照明委员会推荐的城市夜景照度标准

**Illuminance standard value by CIE**
**CIE国际照明委员会推荐的照度标准值**

| Rogues' Gallery materials<br>被照面材料 | Recommended illumination (LX)<br>推荐照度(lx) | | | Correction factor<br>修正系数 | | | | |
|---|---|---|---|---|---|---|---|---|
| | Background brightness<br>背景亮度 | | | Type of light source correction Surface conditions amended<br>光源种类修正 | | Surface conditions amended<br>表面状况修正 | | |
| | Low<br>低 | Middle<br>中 | High<br>高 | Mercury lamp Metal halide lamps<br>汞灯、金属卤化物灯 | High and low pressure sodium lamps<br>高、低压钠灯 | Cleaner<br>较清洁 | Dirty<br>脏 | Very dirty<br>很脏 |
| Light color marble, white marble<br>浅色石材、白色大理石 | 20 | 30 | 60 | 1 | 0.9 | 3 | 5 | 10 |
| Middle color stone, cement, light marble<br>中色石材、水泥、浅色大理石 | 40 | 60 | 120 | 1.1 | 1 | 2.5 | 5 | 8 |
| Dark stone, grey granite<br>深色石材、灰色花岗 | 100 | 150 | 300 | 1 | 1.1 | 2 | 3 | 5 |
| Light yellow brick material<br>浅黄色砖材 | 30 | 50 | 100 | 1.2 | 0.9 | 2.5 | 5 | 8 |
| Light brown brick material<br>浅棕色砖材 | 40 | 60 | 120 | 1.2 | 0.9 | 2 | 4 | 7 |
| Material of dark brown brick, pink granite<br>深棕色砖材、粉红色花岗石 | 55 | 80 | 160 | 1.3 | 1 | 2 | 4 | 6 |
| Red brick<br>红砖 | 100 | 150 | 300 | 1.3 | 1 | 2 | 4 | 5 |
| Dark brick<br>深色砖 | 120 | 180 | 360 | 1.3 | 1.2 | 1.5 | 2 | 3 |
| Architectural concrete<br>建筑混凝土 | 60 | 100 | 200 | 1.3 | 1.2 | 1.5 | 2 | 3 |
| Natural aluminum (surface coatings)<br>天然铝材(表面烘漆处理) | 200 | 300 | 600 | 1.2 | 1 | 1.5 | 2 | 2.5 |
| Reflectivity 10% the dark surface material<br>反射率10%的深色面材 | 120 | 180 | 360 | —— | —— | 1.5 | 2 | 2.5 |
| Reflection rate of 30% per cent of the surface material<br>反射率30%-40%的中色面材 | 40 | 60 | 120 | —— | —— | 2 | 4 | 7 |
| Reflection rate of 60% per cent pink surface material<br>反射率60%~70%的粉色面材 | 20 | 30 | 60 | —— | —— | 3 | 5 | 10 |

(2) Illumination standard value recommended by IESNA(lx)
北美照明学会(IESNA)推荐的照明标准(lx)

| Applications<br>应用场合 | City<br>城市 | Suburb<br>郊区 | Village<br>乡村 |
|---|---|---|---|
| Light marble and plaster<br>浅色大理石和石膏 | 150 | 100 | 50 |

To be Continued 续前表

| Light yellow limestone, smooth and pale yellow brick 浅黄色石灰石、光滑浅黄色砖等 | 200 | 150 | 100 |
|---|---|---|---|
| Smooth black ash grey bricks, brick tiles 光滑灰砖、砖黑灰砖 | 300 | 200 | 150 |
| Brownstone building, other black surface 赤褐色砂石建筑、其他黑色表面 | 500 | 350 | 200 |

(3) Japan illumination standard value by night view guide (lx)
日本城市夜景照明指南推荐照度标准(lx)

| Surface materials 表面材料 | Background brightness 背景亮度 | | bright 亮 | genera 一般 | dark 暗 |
|---|---|---|---|---|---|
| | | | 市中心 | 小城市 | 乡村 |
| | Illumination 照度 | Emission ratio 发射比(%) | 12(cd/m²) | 6(cd/m²) | 4(cd/m²) |
| White marble 白色大理石 | White 白 | 80 | 150 | 100 | 50 |
| Concrete 混凝土 | Bright 明 | 60 | 200 | 150 | 100 |
| Yellow-brown bricks 黄茶色砖 | Middle 中 | 35 | 300 | 200 | 150 |
| Dark grey brick 暗灰色砖 | Dark 暗 | 15 | 500 | 300 | 200 |

Note：bright：Commercial Street, billboard-intensive areas; general：Commercial Street with less billboard; dark：no billboard
注：亮：商业街，广告牌密集地带；一般：广告牌少的商业街；暗：无广告场所；

(4) Germany illumination standard value by night view (lx)
德国城市夜景照明的照度标准(lx)

| Rogues' Gallery materials 被照面材料 | Lighting condition 被照明状况 | surface illumination 被照表面的照度 | | |
|---|---|---|---|---|
| | | Environment Lighting condition 环境照明状况 | | |
| | | Dark 暗 | Middle 中等 | Bright 明亮 |
| White marble 白色大理石 | Very clean 很清洁 | 25 | 50 | 100 |
| Light concrete 浅色混凝土 | Very clean 很清洁 | 50 | 100 | 200 |
| White tile 白色面砖 | Very clean 很清洁 | 20 | 40 | 80 |
| Yellow tile 黄色面砖 | Very clean 很清洁 | 50 | 100 | 200 |
| White granite 白色花岗岩 | Very clean 很清洁 | 150 | 300 | 600 |

To be Continued 续前表

| Concrete or dark stone 混凝土或深色石材 | Very clean 很清洁 | 75 | 150 | 300 |
|---|---|---|---|---|
| Red brick 红砖 | Very clean 很清洁 | 75 | 150 | 300 |
| Concrete 混凝土 | Very dirty 很脏 | 150 | 300 | 600 |
| Red brick 红砖 | dirty 脏 | Minimum value 最低值 | Minimum value 最低值 | Minimum value 最低值 |

**(5) Netherlands illumination standard value by night view (lx)**
荷兰城市夜景照明的照度标准(lx)

| Rogues' Gallery materials 被照面材料 | Environment 环境 | | |
|---|---|---|---|
| | Condition 条件 | Dark 光线暗 | General 光线一般 | Bright 光线明亮 |
| White tile, marble 白色砖、大理石 | Very clean 相当清洁 | 25 | 50 | 100 |
| Light concrete, stone, yellow brick 浅色混凝土、石料、黄色砖 | Very clean 相当清洁 | 50 | 100 | 200 |
| Dark concrete, stone, brick 深色混凝土、石料、红砖 | Very clean 相当清洁 | 75 | 150 | 300 |
| Granite 花岗岩 | Very clean 相当清洁 | 100 | 200 | 400 |
| Red brick 红砖 | dirty 脏 | 150 | 300 | —— |

**(6) China illumination of urban night view lighting standards (lx)**
中国城市夜景照明的照度标准(lx)

| Rogues' Gallery materials 被照面材料 | City size 城市规模 | Average illumination 平均照度 | | | |
|---|---|---|---|---|---|
| | | E1区 | E2区 | E3区 | E4区 |
| White wall of exterior wall paint, white glazed, light cold or warm exterior wall paint, white marble 白色外墙涂料、乳白色外墙釉面砖、浅冷、暖色外墙涂料、白色大理石 | 大 | —— | 30 | 50 | 150 |
| | 中 | —— | 20 | 30 | 100 |
| | 小 | —— | 15 | 20 | 75 |
| Aluminum plastic plates, light silver or grey-green marble, white marble, ceramic tile, grey or light Khaki, medium-light paint, glazed aluminums composite panel 银色或灰绿色铝塑板、浅色大理石、白色石材、浅色瓷砖、灰色或土黄色釉面砖、中等浅色涂料、铝塑板等 | 大 | —— | 50 | 75 | 200 |
| | 中 | —— | 30 | 50 | 150 |
| | 小 | —— | 20 | 30 | 100 |
| Dark natural granite, marble, ceramic tile, concrete, Brown, dark red, artificial granite, glazed brick 深色天然花岗石、大理石、瓷砖、混凝土、褐色、暗红色釉面砖、人造花岗石、普通砖等 | 大 | —— | 75 | 150 | 300 |
| | 中 | —— | 50 | 100 | 250 |
| | 小 | —— | 30 | 75 | 200 |

Note: 1. E1 (a dark environment) environment, building facades should not set lighting.
2. Structure and special landscape elements such as bridges, sculpture, tower, monument, night view of the city walls; without prejudice to the use of functions under the premise showing its form of beauty, and should be adapted to the environment.

注：1. E1区为（天然暗环境区）生态环境，建筑立面不应该设夜景照明。
2. 构筑物和特殊景观元素如桥梁、雕塑、塔、碑、城墙等的夜景，在不影响其使用功能的前提下，展现其形态美感，并应该与环境协调。

## 四、Lighting and energy saving——Green lighting
### 照明与节能——绿色照明

Green lighting is by enhancing the efficiency of lighting equipment and systems, energy conservation; reducing power plant emissions of air pollutants and greenhouse gases, protect the environment; improving the quality of life and work efficiency, and create a culture of modern civilization. Energy wasting is a worldwide problem, mainly in the field of lighting design, the energy loss at the same time is resulting in light pollution. The world will focus on conventional energy resources and the development of new energy saving, countries are committed to lighting energy-saving standard of research work, the following is nations' green lighting standard.

绿色照明是指通过提高照明电器和系统的效率，节约能源；减少发电排放的大气污染物和温室气体，保护环境，改善生活质量，提高工作效率，营造体现现代文明的光文化。能源浪费问题是一个世界性的问题，它在照明设计领域主要体现为大量的能源流失的同时，产生光污染。目前全球都将目光放在了节约传统能源和开发新能源方面，各个国家都在致力照明节能标准的研究工作，以下是各国的绿色照明标准。

(一) America lighting energy saving standards：
美国照明节能标准：

United States in 1990s began working on lighting energy-saving standard, and each State has a corresponding energy consumption limits standard, including: residential, Office, commercial, sports, transportation, hospitals, schools and other buildings.

美国九十年代就开始致力于照明节能标准的研究工作，并且各州都有相应的能耗限制标准。而且包括了：居住、办公、商业、体育、交通运输、医院、学校等建筑。

United States green lighting light standards
美国绿色照明的照度标准

| Building or area types<br>建筑或区域类型 | The entire building<br>整栋建筑<br>W/m² (W/ft²) | Part of the building or room<br>房间或建筑-部分<br>W/m² (W/ft²) |
|---|---|---|
| Auditorium 礼堂 | —— | 19.38(1.8) |
| Automotive facilities 汽车设施 | 9.69(0.9) | —— |
| Banks/financial institutions 银行/金融机构 | —— | 16.15(1.5) |
| Classroom/lecture hall 教室/讲堂 | —— | 15.07(1.4) |
| Conference rooms, Congress or Conference Center<br>会议厅、代表会议或会议中心 | 12.92(1.2) | 13.99(1.3) |
| Corridors, lounges, a secondary zone<br>走廊、休息室、辅助区 | —— | 9.69(0.9) |
| Cathedral/City Hall 教堂/市政厅 | 12.92(1.2) | —— |
| Dining 就餐 | —— | 9.69(0.9) |
| Hostel 宿舍 | 10.76(1.0) | —— |
| Training Center 训练中心 | 10.76(1.0) | 9.69(0.9) |
| Exhibition Center 展览中心 | —— | 13.99(1.3) |
| Grocery stores 杂货店 | 16.15(1.5) | 17.22(1.6) |
| Sports Centre playground 体育馆运动场地 | —— | 15.07(1.4) |
| Hotel 旅馆 | 10.76(1.0) | 13.99(1.3) |
| Industrial production, < 20 feet ceiling height<br>工业生产，< 20英尺顶棚高 | —— | 12.92(1.2) |
| Industrial production, ≥ 20 feet ceiling height<br>工业生产，≥20英尺顶棚高 | —— | 18.30(1.7) |
| Kitchen 厨房 | —— | 12.92(1.2) |
| Reading room 阅览室 | 13.99(1.3) | 18.30(1.7) |

To be Continued 续前表

| | | |
|---|---|---|
| Hotel lobby 旅馆大堂 | —— | 11.84(1.1) |
| Other halls 其他厅堂 | —— | 13.99(1.3) |
| Shopping Center, arcade of shops on both sides, vestibular 购物中心、拱廊两侧商店、前庭 | —— | 6.46(0.6) |
| Medical and health care 医疗和医疗保健 | 12.92(1.2) | 12.92(1.2) |
| Motel 汽车旅馆 | 10.76(1.0) | —— |
| Apartments 公寓 | 7.53(0.7) | —— |
| Museum 博物馆 | 11.84(1.1) | 10.76(1.0) |
| Office 办公室 | 10.76(1.0) | 11.84(1.1) |
| Garage 车库 | 3.23(0.3) | —— |
| Prison 监狱 | 10.76(1.0) | —— |
| Police station/fire station 公安派出所/消防站 | 10.76(1.0) | —— |
| Post Office 邮政局 | 11.84(1.1) | —— |
| Religious ceremony Hall 宗教仪式堂 | 13.99(1.3) | 25.83(2.4) |
| Restaurant 餐馆 | 17.22(1.6) | 9.69(0.9) |
| Retail and wholesale showroom 零售、批发展示厅 | 16.15(1.5) | 18.30(1.7) |
| School 学校 | 12.92(1.2) | —— |
| Industrial and commercial warehouses 工业和商业仓库 | 8.61(0.8) | 8.61(0.8) |
| Cinema 电影院 | 12.92(1.2) | 12.92(1.2) |
| Theatre 剧院 | 17.22(1.6) | 27.99(2.6) |
| Traffic and transport 交通运输 | 11.84(1.1) | —— |
| Others 其他 | 6.46(0.6) | 10.76(1.0) |

(二) Japan lighting energy saving standards:
　　日本照明节能标准:

Japan set out in the "law on energy" about hotels, offices, hospitals and clinics, schools, shops, restaurants, six categories of building lighting power density ($W_s$).
日本《节能法》中规定了饭店旅馆、办公、医院和诊所、学校、商店、餐饮店等六类建筑的照明功率密度($W_s$)。

| | Hotels 饭店和旅馆 | | | |
|---|---|---|---|---|
| Category 类别 | District 区域 | Examples of space objects 空间对象举例 | | $W_s$ (W/m²) |
| 1 | Public space 公共空间 | Entrance, restaurant, Banquet Hall 入口、餐厅、宴会厅 | | 30 |

| Category | District | Examples of space objects空间对象举例 | $W_s$ (W/m²) |
|---|---|---|---|
| 2 | Room, Application a<br>客房、应用a | Hall, Office of passenger escalator, shop, kitchen, Office, conference room, preparation room, Center for disaster prevention, management rooms, control rooms<br>大堂、客用扶梯厅、店铺、厨房、办公室、会议室、准备室、防灾中心、管理人员室、控制室 | 20 |
| | | Guest room, guest bathroom, dressing rooms, control rooms<br>客房、客用卫生间、更衣室、控制室 | 15 |
| 3 | Application b<br>应用b | Linen, warehouses, corridors, rooms, staff, room with bathroom, staff changing rooms, a staff with a channel, staff with stairs<br>布巾室、仓库、走廊、工作人员室、工作人员用卫生间、工作人员用更衣室、工作人员用通道、工作人员用楼梯 | 10 |
| 4 | Application b (machine room)<br>应用b(机械室等) | Machine room, electric room, warehouse, others<br>机械室、电气室、仓库、其它 | 5 |

### Office building 办公楼

| Category类别 | District区域 | Examples of space objects空间对象举例 | $W_s$ (W/m²) |
|---|---|---|---|
| 1 | Public space, fine visual task<br>公共空间、非常精细的视觉作业 | Entrance room, operating room, design<br>入口、营业室、设计室 | 30 |
| 2 | Office, Application a<br>办公室空间应用a | Office, staff room, conference room, VDT/CAD room, bathrooms, dressing rooms, elevators, disaster prevention centre, room management, monitoring room, control room<br>办公室、工作人员室、会议室、VDT/CAD室、卫生间、更衣室、电梯厅、防灾中心、管理人员室、监视室、控制室 | 20 |
| 3 | Application b<br>应用b | Boiler Room, corridors, passageways<br>开水房、走廊、通道 | 10 |
| 4 | Application b (machine room)<br>应用b(机械室等) | Machine room, electric room, storage, stairs, driveways, parking lots, other<br>机械室、电气室、仓库、楼梯、车道、停车场、其它 | 5 |

### Hospitals and clinics 医院和诊疗所

| Category类别 | District区域 | Examples of space objects空间对象举例 | $W_s$ (W/m²) |
|---|---|---|---|
| 1 | fine visual task<br>非常精细的视觉作业 | 手术室 | 55 |
| | | 诊疗室、药房 | 30 |
| 2 | Treatment room Application a<br>处置室、应用a | Lobby, registration, elevator room, examination room, treatment rooms, set in the treatment room, preparation room, nurses stations, offices, conference rooms, a disaster prevention Center, room management, monitoring room, control room<br>大堂、挂号、电梯厅、检查室、处置室、集中治疗室、准备室、护士站、办公室、会议室、防灾中心、管理人员室、监视室、控制室 | 20 |
| | | document room, waiting room, dining room, kitchen, store, bathroom, lounge, a guards' room, changing rooms<br>资料室、候诊室、食堂、厨房、小卖部、卫生间、休息室、值班室、更衣室 | 15 |
| 3 | Ward、Application b<br>病房、应用b | Ward, linen equipment rooms, corridors, stairways<br>病房、布巾器材室、走廊、楼梯 | 10 |
| 4 | Application b (machine room)<br>应用b(机械室等) | Machine room, electric room, warehouse, driveways, parking lots, other<br>机械室、电气室、仓库、车道、停车场、其它 | 5 |

| Schools学校 | | | |
|---|---|---|---|
| Category类别 | District区域 | Examples of space objects空间对象举例 | $W_s$ (W/m²) |
| 1 | Lecture hall、fine visual task 讲堂、非常精细视觉作业 | 讲堂、计算机教室 | 30 |
| 2 | classroom Application a 教室、应用a | Classrooms, lecture rooms, laboratories, a gymnasium, staff room, meeting room 教室、讲义室、实验室、体育馆、职员室、会议室 | 20 |
| | | Libraries, laboratories, offices, radio room, logistics officer 图书室、实习室、办公室、广播室、后勤人员室 | 15 |
| 3 | Application b 应用b | Dining room, kitchen, dressing room, bathroom, corridor 食堂、厨房、更衣室、卫生间、走廊 | 10 |
| 4 | Application b (machine room) 应用b(机械室等) | Machine room, electric room, warehouse, other 机械室、电气室、仓库、其它 | 5 |

| Shops商店 | | | |
|---|---|---|---|
| Category类别 | District区域 | Examples of space objects空间对象举例 | $W_s$ (W/m²) |
| 1 | Public space 公共空间 | Entrance 入口 | 55 |
| | | Passenger elevator 客用电梯 | 30 |
| 2 | Business Hall Application a 营业厅、应用a | Hall, guest channels 营业厅、客用通道 | 20 |
| | | Dining room, guest bathroom, a disaster prevention Center, room management, monitoring room, control room 餐厅、客用卫生间、防灾中心、管理人员室、监视室、控制室 | 15 |
| 3 | Application b应用b | Passenger stairs, back part of the room, staff, staff changing rooms, with bathroom, staff of staff channel 客用楼梯、后部分、工作人员室、工作人员更衣室、工作人员用卫生间、工作人员用通道 | 10 |
| 4 | Application b (machine room) 应用b(机械室等) | Mechanical electrical rooms, warehouses, staff room, stairs, driveways, parking 机械室、电气室、仓库、工作人员用楼梯、车道、停车场其它 | 5 |

| Restaurants餐饮店 | | | |
|---|---|---|---|
| Category类别 | District区域 | Examples of space objects空间对象举例 | $W_s$ (W/m²) |
| 1 | Public space 公共空间 | The central part (stairs), the population of passenger elevator Hall, a variety of shops, guest channels 中央部分（扶梯）、客用电梯厅、各种店铺入口、客用通道 | 30 |
| 2 | Seat Application a 座位、应用a | Seating, registration, kitchen, offices, and disaster prevention Center, room management, monitoring room, control room 坐席、登记、厨房、办公室、防灾中心、管理人员室、监视室、控制室 | 20 |
| | | Waiting room, guest bathroom, guest access, passenger stairs 等候室、客用卫生间、客用通道、客用楼梯 | 15 |

To be Continued 续前表

| | | | |
|---|---|---|---|
| 3 | Tea Application b<br>喝茶、应用b | Tea House seat toilet, staff rooms, staff changing rooms, staff, staff with stairs<br>茶馆的坐席、工作人员室、工作人员更衣室、工作人员用卫生间、工作人员用楼梯 | 10 |
| 4 | Application b (machine room)<br>应用b（机械室等） | Machine room, electric room, warehouse<br>机械室、电气室、仓库 | 5 |

## (三)Russia lighting energy saving standards
俄罗斯照明节能标准

| Room name<br>房间名称 | According to MTCH "natural day lighting and artificial lighting" standard of illumination (LX)<br>根据MTCH"天然采光和人工照明"的照度标准(LX) | Maximum allowed units installed power(W/m²)<br>最大允许单位安装功率(W/m²) |
|---|---|---|
| Administration building (State departments, Chief departments, committees, management, and so on), design organizations, scientific research institutes and libraries<br>管理建筑（国家部门的、主管部门的、委员会的、管理部门的等），设计机构，科研机关和图书馆等 | | |
| (1)laboratory and Studio, Office and so on;<br>研究室和工作室、办公室等 | 400 | 25 |
| (2)design room and design office, designing and drawing rooms<br>设计室和设计厅、设计和制图室 | 500 | 35 |
| (3)the copy room and Studio; photo room<br>复印室和照相室等； | 400 | 25 |
| (4)file, Visual terminals, monitors workplaces<br>有文件、视觉终端、监视器等工作间 | 400 | 25 |
| (5)Reading room<br>阅览室 | 400 | 25 |
| (6)Laboratory<br>实验室 | 500 | 35 |
| Banks and insurance agencies: Business Office, cash office<br>银行和保险机构：营业厅、现金出纳厅 | 500 | 35 |
| General secondary and primary schools, boarding schools, vocational and technical schools, secondary specialized and higher education: teaching classrooms, Lecture Hall, staff room, laboratory, Laboratory Assistant, information and technology<br>普通中、小学、寄宿学校、职业技术学校、中等专科和高等学校：教室、讲堂、教学人员室、实验室、实验员的房间、信息和计算技术室 | 400 | 25 |
| Preschool children: the group room, game room, dining room, music and physical education classrooms.<br>学龄前儿童机构：小组教室、游戏室、用餐室、音乐和体育课教室 | 400 | 25 |
| Public catering Enterprise: Hall, Snack room, Tea room;<br>公共饮食企业：营业厅、小吃部、茶点部 | 200 | 14 |
| Western-style food dining house<br>西餐间 | 400 | 25 |
| Store：Supermarket sales lobby<br>商店：超级市场的售货大厅 | 35 | 500 |

To be Continued 续前表

| | | |
|---|---|---|
| Store sales hall<br>商店的售货厅 | 25 | 400 |
| Residential services enterprises: hair salon<br>居民生活服务企业：理发室 | 400 | 35 |
| Sewing workshop and clothing repair shop<br>缝纫车间和衣服修理部 | 750 | 52 |
| Pharmacy: Customer Service Office<br>药房：顾客服务厅 | 200 | 14 |
| Residents of building: bedroom<br>居民建筑：寝室 | 300 | 20 |
| Residential outdoor tiers corridors, stairwells and lobby<br>住宅各户外的各层走廊、楼梯间和门厅 | 30 | 4 |
| Indoor car park, warehouse: Indoor parking of transport enterprises and public agencies, transportation vehicles room<br>室内停车场、库房：室内停放运输企业和公共机构的交通运输车辆的房间 | 75 | 10 |
| Transportation enterprise technical service stations；bus stop<br>交通运输技术服务站交通运输企业；汽车站 | 200 | 14 |
| Vehicle inspection station<br>汽车检验站 | 300 | 20 |
| Technical service stations<br>技术服务站 | 200 | 14 |

## (四)China lighting energy saving standards
### 中国照明节能标准

China's lighting energy saving standards has two kinds, national standard of "architectural lighting design standard" (GB50034-2004), and local standards, such as: Beijing standard green lighting project technical regulations (DBJ01-607-2001), the Shanghai local standard "lighting equipment and reasonable electricity standard" (DB 31/178-1996). The following introduces a national standard GB50034-2004 energy saving standards.

中国的照明节能标准有国家标准《建筑照明设计标准》(GB50034—2004)，还有地方标准如：北京市标准《绿色照明工程技术规程》(DBJ01—607—2001)，上海市地方标准《照明设备合理用电标准》(DB 31／178—1996)。以下主要介绍国家标准GB50034—2004的节能标准。

| Category类别 | Lighting power density(W/m$^2$)<br>照明功率密度(W/m$^2$) | |
|---|---|---|
| | Present现行值 | Target目标值 |
| Residential building per unit<br>居住建筑每户 | 7 | 9 |
| Office building<br>办公建筑 | | |
| Office building<br>普通办公室 | 11 | 9 |
| High-grade office<br>高档办公室 | 18 | 15 |
| Conference Room<br>会议室 | 11 | 9 |
| Business Hall<br>营业厅 | 13 | 11 |
| Documentation, copy, release room<br>文件整理、复印、发行室 | 11 | 9 |

| | | |
|---|---|---|
| Registry room<br>档案室 | 8 | 7 |
| Commercial buildings<br>商业建筑 | | |
| The general store business hall<br>一般商场营业厅 | 12 | 10 |
| High-end stores are open hall<br>高档商店营业厅 | 19 | 16 |
| General supermarket business hall<br>一般超市营业厅 | 13 | 11 |
| High-end supermarket business hall<br>高档超市营业厅 | 20 | 17 |
| Hotel rooms<br>旅馆客房 | 15 | 13 |
| Hotel Chinese restaurant<br>旅馆中餐厅 | 13 | 11 |
| Hotel multi-function hall<br>旅馆多功能厅 | 18 | 15 |
| Hotel room corridor<br>旅馆客房层走廊 | 5 | 4 |
| Hotel lobby<br>旅馆门厅 | 15 | 13 |
| Hospital buildings<br>医院建筑 | | |
| Treatment room, clinic<br>治疗室、诊室 | 11 | 9 |
| Laboratory<br>实验室 | 18 | 15 |
| Operating room<br>手术室 | 30 | 25 |
| Waiting room, registration hall<br>候诊室、挂号厅 | 8 | 7 |
| Ward<br>病房 | 6 | 5 |
| Nurses' station<br>护士站 | 11 | 9 |
| Pharmacy<br>药房 | 20 | 17 |
| Intensive care room<br>重症监护房 | 11 | 9 |
| School buildings<br>学校建筑 | | |
| Classrooms, reading room<br>教室、阅览室 | 11 | 9 |
| Laboratory<br>实验室 | 11 | 9 |
| The art room<br>美术教室 | 18 | 15 |
| Multimedia classrooms<br>多媒体教室 | 11 | 9 |

# 01 Office&Cultural
## 办公+文化

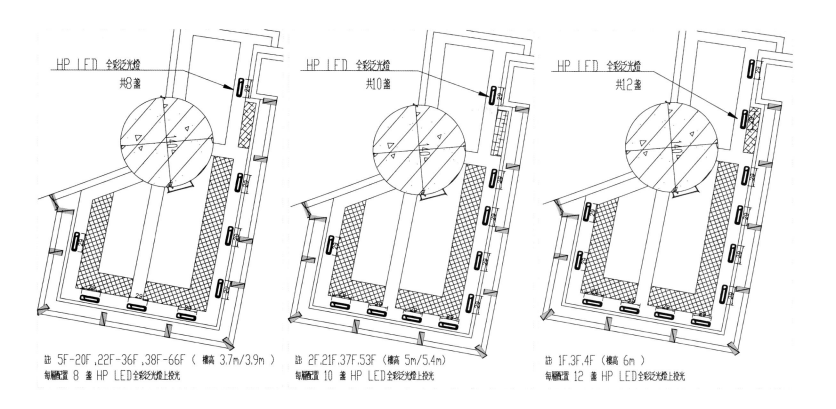

- ──155 North Wacker Drive
  芝加哥北威克街155号

  The strong geometric forms and integrated lighting solutions contribute to the success of this exciting new public space.
  清晰的几何造型和综合性的照明措施成就了这座热情四溢的建筑。

- ──Old DC Courthouse
  华盛顿哥伦比亚特区法院

  Mix in the need to maintain a low-energy profile, and the expectation to retain historic luminaires; add budget and time constraints; and a complex, yet rewarding, project can emerge.
  维护低能耗设备，保留历史性灯具，增加预算，突破并拉长工期：由此诞生了一个复杂却富于回报的项目。

- ──"Lightstream" Dallas Convention Center
  达拉斯会展中心

  A river of light glows overhead. Streams of color pulse in waves of illumination. Floating lines in space suggest surface and depth.
  光河流淌在空中，色彩的溪流涌动在照明波浪里。空间里漂浮的线条框出了表面和深度。

- ──Roca Barcelona Gallery
  乐家巴塞罗那展厅

  The use of artificial light on the glass facade at dusk acts as a veil, and creates a sensations of fluidity, tranquility, smooth dynamism, an almost hipnotic mood, to both the exterior and interior spaces.
  玻璃幕墙的人造灯饰，在黄昏里俨然一款面纱，流动、静谧、畅快、动感，这种催眠的状态弥漫于整个室内和室外空间。

- ──Pio Palace in Carpi
  卡尔皮皮奥宫殿

  The principal object of our design was to include elements that were not invasive in terms of aesthetics and conservation of the rooms, with frescos completely covering both walls and ceiling.
  设计的主旨在于，所融入的元素不能体现出侵略性，如美学、房间的保护、墙面和天花板的整体壁画等。

- ──Fondazione Arnaldo Pomodoro, Milan
  米兰阿纳尔多·波莫多罗基金会展览馆

  The most proper lighting of a painting must, instead, transfer coherently what the author wants to tell.
  最适合画的照明一定是传递了画家想要表达的内容。

# TBA Tower

## 夜晚中的生命质感
——TBA环球经贸中心

**Credits**
Location: Dongguan, Guangdong, China
Total Gross Floor Area: 280,000 m²
Client: Dongguan Jinmao Development
Lighting Design: GUANG Architecture Lighting Design
Architect/ Engineer: Shanghai Archiman Architectural Design, ARUP
Photography: GUANG Architecture Lighting Design
Light Source: CDM, LED
Honours: LEED-CS Gold Certification
Cost: RMB14,800,000 (USD$2,333,800) (Lighting only)

Le Corbusier said Light creates ambience and feel of a place, as well as the expression of a structure. Light is one of the most vital natural elements. The world, shape, size, and colour can be understood because of the light and it deeply affects people's ability to see, also ability to feel.

TBA Tower, which is a unique project, has been constructed since 2009 in Dongguan where is one of the major economical developing cities in Mainland China. The tower is a high-rise building which is 289m high and located in the centre of commercial area in Dongguan. The land covers 27,000m², and building has 72 stories of which measure is approximately 280,000m². Based on three scales, the lighting scheme simplifies the colours, and applies soft light sources to make the building smoothly integrating with city and resident life at night. The concept aims to create a combination of both senses of movement and still within one building by illuminations. And, the design collaborates with Artman Architecture

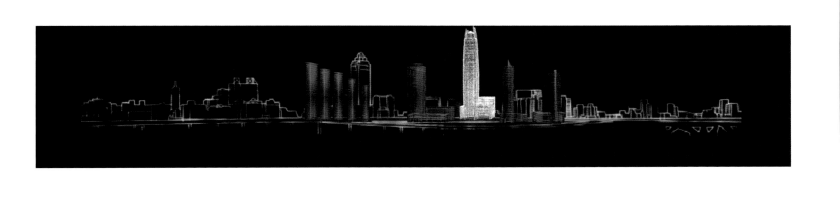

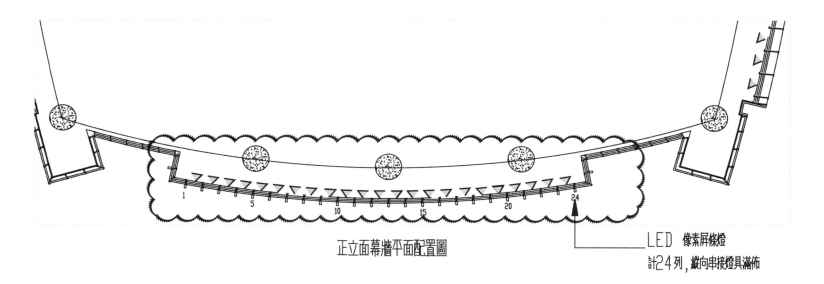

正立面幕牆平面配置圖

LED 像素屏條燈
計24列，縱向串接燈具滿佈

Design Studio and includes building façade and landscape architecture.

### Flowing Light

Unlike another type of façade lighting design which creates illuminations from outside of building, the key idea tends to well conceal lighting as part of architecture structure, so that the light can appear and fade inside out. The building has elegant vertical arch shape, steel construction and enveloped by glass panels. As well as the theme of flowing, lights on different areas of the building will change colours constantly and shift slowly through entire building which is managed by the new smart lighting control system.

The lighting scheme on façade has five main areas: foundation area, glass panels, corner light box, tower top and interactive LED screen. Different from the upper level, tower's foundation is formed with rectangular shape and applies combination light sources, 15W High Power LED and CDM luminaires,

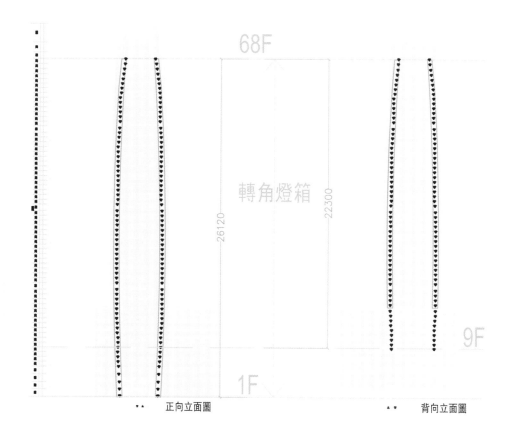

轉角燈箱燈具立面配置圖

正向立面圖　　　背向立面圖

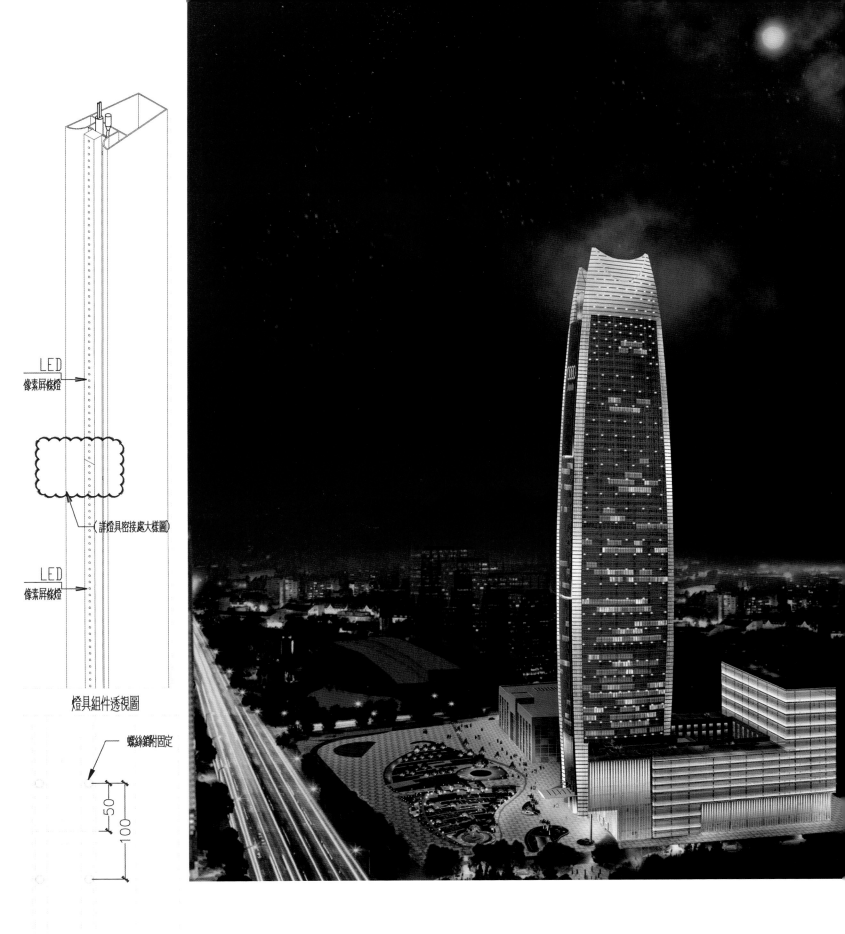

LED
像素屏條燈

(詳燈具密接處大樣圖)

LED
像素屏條燈

燈具組件透視圖

螺絲鎖附固定

50
100

燈具密接處大樣圖

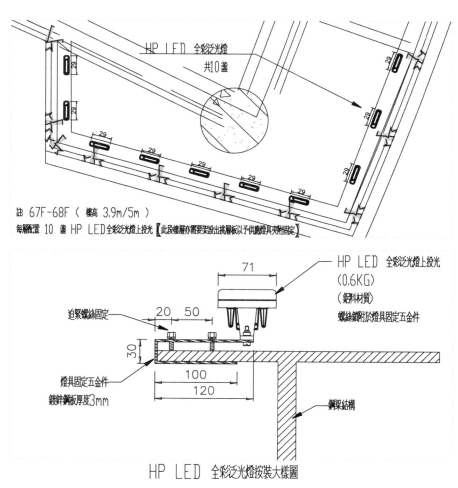

誌 67F-68F（樓高 3.9m/5m）
每層配置 10 盞 HP LED 全彩泛光燈上投光【此段樓層亦需要架設出挑層板以予供應燈具次附固定】

HP LED 全彩泛光燈按裝大樣圖

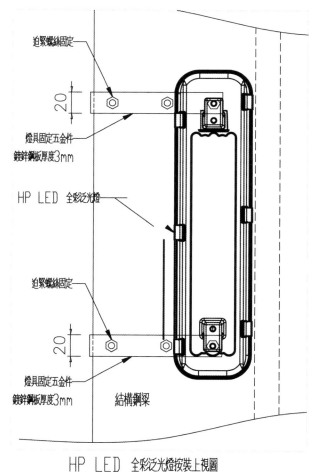

HP LED 全彩泛光燈按裝上視圖

vertically and horizontally on steel structure elements to emphasize the shape of building. Glass panels at both sides of the façade lay upon one of each as scale. And, the idea here is to make illuminations sparkle slowly and randomly as star in the sky. For the effect, nearly 270 4.5W RGB LED linear luminaires are installed separately and hided at the edge of glass panels.

Different from the lighting on the scale, four building corners have concealed over 1,580 25W Colour-change LED luminaires on internal beams and floodlight the space to create light boxes. These installations outline the building's vertical arch shape and create continuous lighting from the ground to the top. The end of light boxes is the top of building which forms like fish mouth and is illuminated by nearly 700 25W High Power RGB LED luminaires. From level 47 to 65, the interactive LED screen is mounted on architectural façade which is total 72m high and consists of over 1,000 luminaires. The screen display is able to be managed by programmable system to present various themes depend on the special events. This idea creates a communicating channel to enhance the interaction between the local business men and residents.

**Static Light**

The lighting scheme for the garden on fifth floor is connecting the nature elements, water and plant, by the illuminations. Instead of using large quantities of luminaires, the concept limits the number of light sources and reduces the lighting to the minimum which aims to create a quiet atmosphere in the

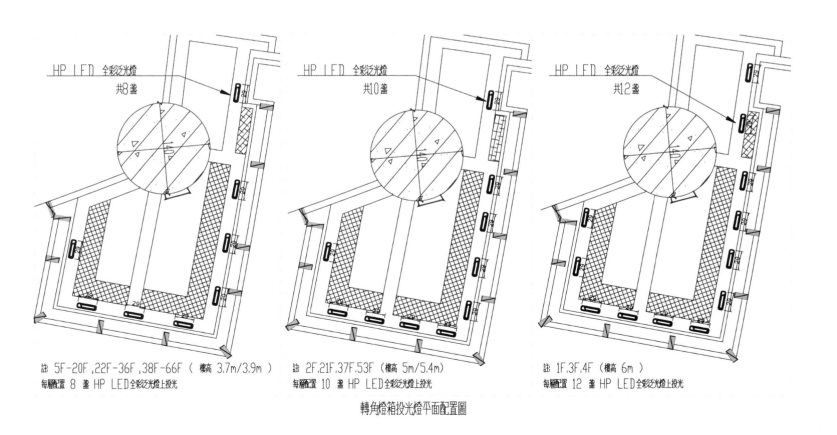

轉角燈箱投光燈平面配置圖

open space. Accent lights are arranged with a rhythm and emphasise part of main garden features. A small waterfall goes across level 6 to level 5 and total 60 50W MR16 are mounted at back of water. The effect creates multiple lighting layers in the water and expresses the life in the garden. An oblong glass dome which is also a sealed sky window is located at level 5. The dome is surrounded by water and allows skylight to go through at daytime. It concealed 20 400W HQI luminaires under glass and is floodlighted softly.

**Intelligent Lighting Control System**

In terms of energy efficiency, the project utilised LED and CDM as major light sources and combined Intelligent lighting control system to manage the light switch on and off. It is a challenge that integrating lighting control system which needs to manipulate and monitor over 10,000 light sources at once. And, the result which is made by those efforts is TBA Tower has achieved LEED-CS gold pre-certification in 2009. The combination of all of innovative designs will lead TBA Tower becoming one of iconic landmarks in Dongguan and create a new image of architecture in the city's night view. It is a cutting edge of lighting design art work which takes various considerations, people, culture, economy, and environment, into account. It expresses a new relation between architecture, lighting, and technology, also brings a novel experience to people.

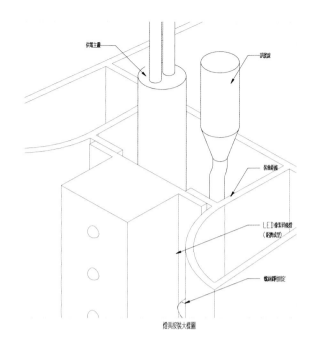
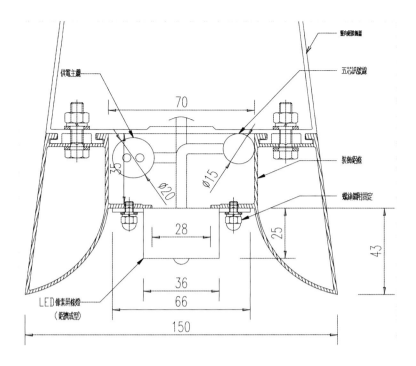
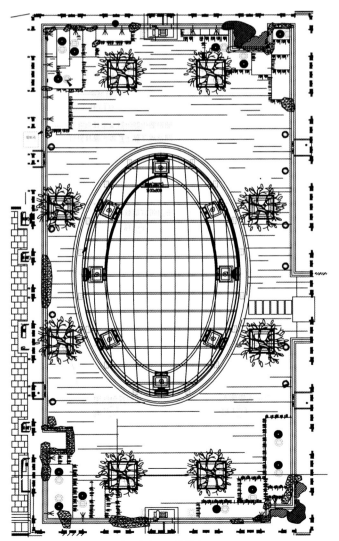

建筑大师勒·柯布西耶曾说"灯光，正如结构一样，赋予建筑特殊的感觉与氛围"。人类视觉架构的感知是透过光，而传递光的技术是照明，人们透过照明设计来感受光的艺术，而照明设计则是运用技术来传递光的价值。

此次与雅门建筑师设计事务所合作建造中的TBA环球经贸中心(台商大厦)，坐落于中国东莞新城市中心的高级金融商业地区东莞大道东侧，为总高289米的高层商业办公大楼，其总用地面积27 000平方米，总建筑面积约280 000平方米、地上68层、地下4层。此照明规划区域包含建筑物及景观整体，设计概念是将照明结合至建筑结构，创造出动态建筑，并搭配景观城市绿洲，展现一动一静的氛围，赋予建筑的夜晚中的生命质感。

### 立面动态图像化建筑

本案照明设计是以基本3尺度(Urban城市-Street街道-Human人行)规划建筑物整体，将建筑立面图像化并融合于东莞的夜景之中。建筑物型态以弧度钢骨为垂直架构，底层为横向几何方体，建筑外层材质以玻璃帷幕打造。照明概念以结合光与建筑结构让光线从建筑物内部渗透出来为主要表现形式。底层空间结合了商场及办公室，以低色温照明配置缓和城市生活步调，越趋高层则以富有科技感的高色温设定为主照明。

"细节"为此项目其中一项挑战，如何将灯具安装隐藏于建筑内部并将建筑结构表达细节化，让外观呈现建筑最简洁利落的样貌为设计原则。建筑底层方形结构运用复合式光源15瓦大功率LED全彩线形灯及复金属CDM投光灯随着结构于立面做直、横向安排，勾勒出材质结构箱体的锐利感。近270支的4.5瓦LED全彩线形点光源配置于建筑两侧立面鱼鳞结构并于夜空中缓慢地闪烁。相较于分散的点光源式安排，建筑四方角落以建构于内部、超过1 580支的25瓦变色LED泛光灯连贯转角灯箱并从底部延续至天际，鱼口灯箱勾画出建筑弧度。顶层鱼口灯箱同样地使用近700支全彩RGB及25瓦大功率LED线型灯点亮建筑高层，而夹于鱼鳞立面之间的建筑立面，则使用上千支的LED条型灯构成巨型建筑立面屏幕，配合照明控制系统的运用，屏幕可于平日或庆典时与观赏者做视觉上的互动。

### 景观宁静的光

位于1楼及5楼的空中花园景观照明概念是以光为主轴，串连造景中的水及植栽。重点式的上投植栽照明加上有节奏的水底布光，运用少量及透过植物后所表现出来的光制造出静谧的氛围。以楼层瀑布连接6楼与5楼空中花

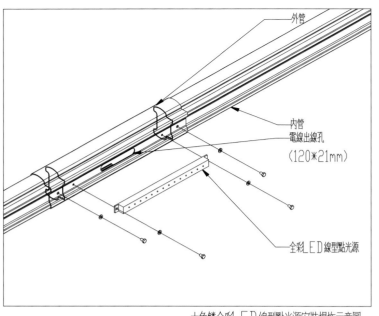

大魚鱗全彩LED線型點光源安裝爆炸示意圖

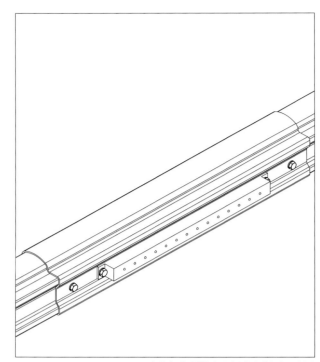

大魚鱗全彩LED線型點光源安裝組合示意圖

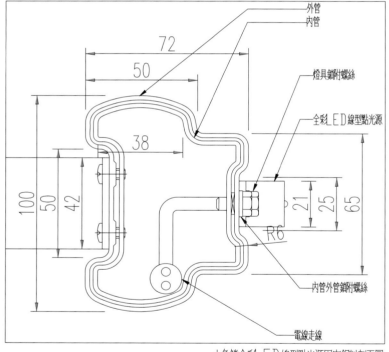

大魚鱗全彩LED線型點光源固定鋁料剖面圖

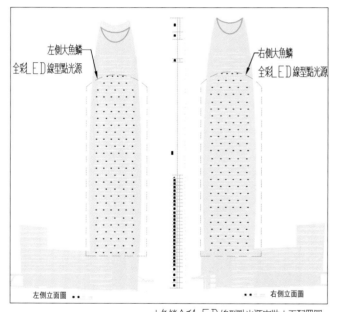

大魚鱗全彩LED線型點光源安裝立面配置圖

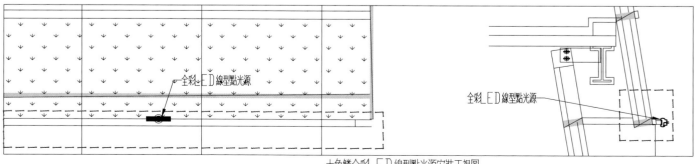

大魚鱗全彩LED線型點光源安裝正視圖

园及水中的建筑穹顶为景观主要特色。于楼层瀑布后方配置50瓦上投水中嵌灯，透过水的延伸创造视觉层次感。建筑穹顶，白天作为室内空间自然采光罩，夜晚则藉由安装在穹顶内部的400瓦投光灯，营造出透露着柔和光线的玻璃圆顶。

照明控制

此项目使用的主要光源为高发光效率LED及CDM，同时配合智能型照明控制器，在各合作单位的精心设计规划下，建筑物已通过LEED-CS金级预认证。TBA建筑照明使用灯具总数量高达10 000组以上，同时必须由中央控制系统控制照明效果并管理能源使用效率，因此控制系统的整合及操作便成为另一项设计的挑战。TBA立面有如呼吸频率般闪烁的鱼鳞点光源、连续缓慢向上流动的转角灯箱、持续可调光发亮的鱼口灯箱、立面巨型屏幕内容的展现、于深夜渐暗的整体亮度，所应用的即是智能型照明控制系统整合后的效果。

在不造成光害的条件下，可调光的光源结合智能型照明的时间控制创造出律动的城市步调及互动性，同时，透过智能型照明监控能源的使用并控制节目安排的编写，不但带给建筑本身经济上的效益，并让居住在与建筑同个城市的人们，可透过照明与建筑互动。其整体设计所呈现的是中国东莞的明日性及未来感，在艺术层面所展现的是一种静与动、建筑及人类互动的结合，于技术面则是应用新颖的软硬件及多元的布光手法，创造出在实践人文艺术理念的同时也尊重环境的创作。

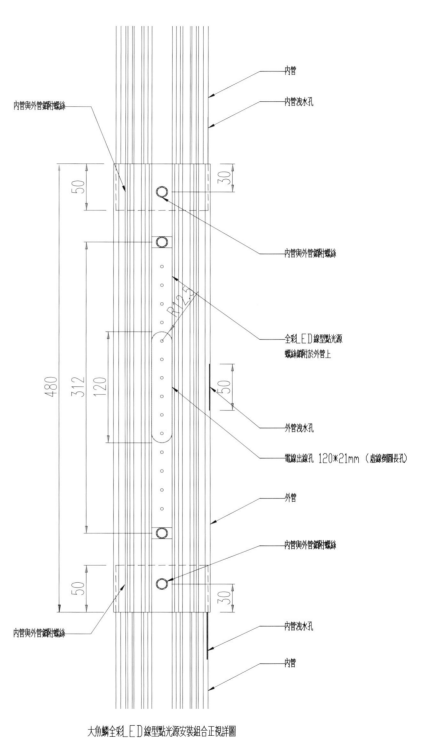

大魚鱗全彩LED線型點光源安裝組合正視詳圖

# 155 North Wacker Drive

高性能，低维护
——芝加哥北威克街155号

**Credits**
Location: Chicago, Illinois, USA
Scale: 46 stories, 104,888 m² Office Tower
Lighting Design: Stephen Margulies, One Lux Studio
Firm: Goettsch Partners
Architects: Jim Goettsch and Steve Nilles
Photography: Tom Rossiter
Light Source: Cove Lights-LED (Color Kinetics); Recessed downlights-Ceramic
Metal Halide (Kurt Versen)
Recognition: 2011 LUMEN Award of Merit IESNA NY Chapter

The new 1.4 million square foot office tower boasts a brand new lobby and arcade that activates the streetscape and creates a dynamic public space entry into this new corporate center. The lobby and arcade are contiguous to each other and a powerful diagonal grid is overlayed to both spaces. The interior lobby and exterior arcade appear seamlessly connected by ceiling and lighting design.

The strong diagonal lines are created using LED strip lights mounted within an architecturally constructed cove. These concealed light sources were used for their ability to perform unaffected by changes in ambient temperature as the cove exists both in interior as well as exterior environments.

Recessed double lamp ceramic metal halide accent lights are located within the cove to provide appropriate ambient and feature light levels. Concealing these fixtures in the cove provides a clean uncluttered ceiling.

The stone core is illuminated with recessed metal halide accent lights fitted with spread lens assemblies to provide a soft wash of light. The perimeter cove uses indirect fluorescent strip lights to float the ceiling from the edge wall. All light sources are color matched to 3,000 degrees K. The elevator lobbies are lighted using asymmetric concealed ceramic metal halide flood lights. These fixtures are concealed in a "carved" slot in the core walls. The indirect lighting provides a warm and inviting effect. The strong geometric forms and integrated lighting solutions contribute to the success of this exciting new public space.

The project has achieved LEED CS Gold rating. All light sources are very efficient with low maintenance. The power density for the lobby was achieved at 1.2 watts/sf.

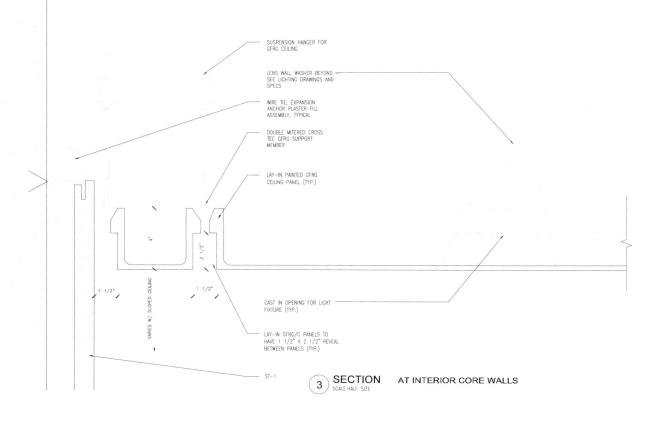

3 SECTION AT INTERIOR CORE WALLS
SCALE HALF SIZE

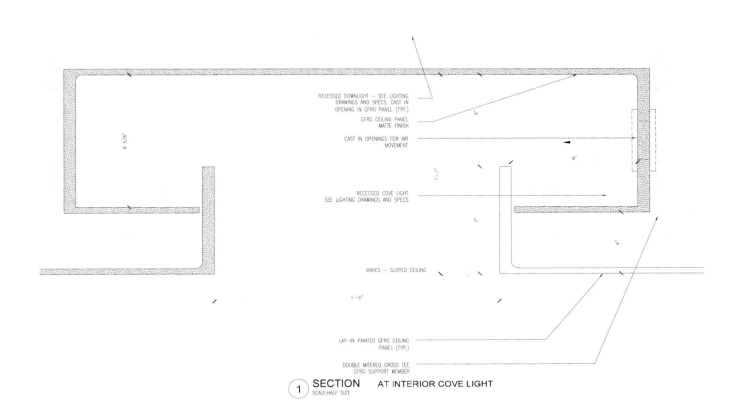

RECESSED DOWNLIGHT – SEE LIGHTING DRAWINGS AND SPECS. CAST IN OPENING IN GFRG PANEL (TYP.)

GFRG CEILING PANEL MATTE FINISH

CAST IN OPENINGS FOR AIR MOVEMENT

RECESSED COVE LIGHT SEE LIGHTING DRAWINGS AND SPECS

VARIES – SLOPED CEILING

LAY-IN PAINTED GFRG CEILING PANEL (TYP.)

DOUBLE MITERED CROSS TEE GFRG SUPPORT MEMBER

**1  SECTION  AT INTERIOR COVE LIGHT**
SCALE: HALF SIZE

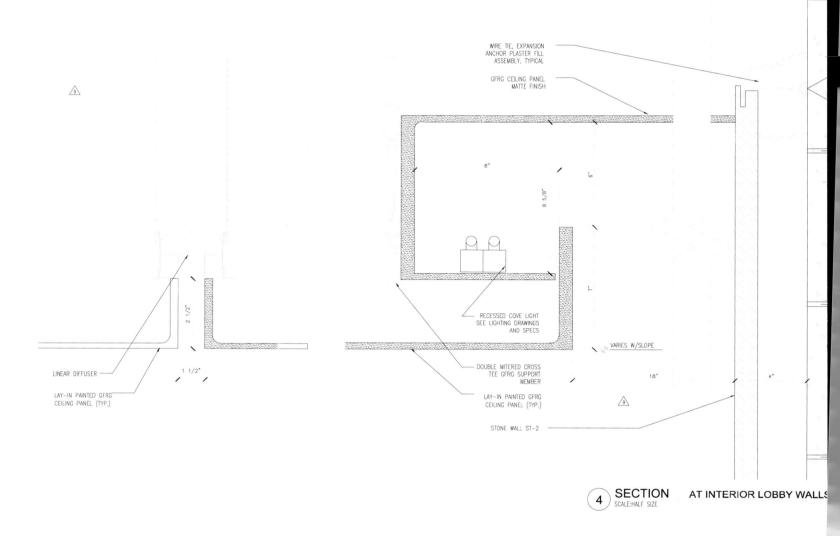

④ **SECTION** AT INTERIOR LOBBY WALLS
SCALE:HALF SIZE

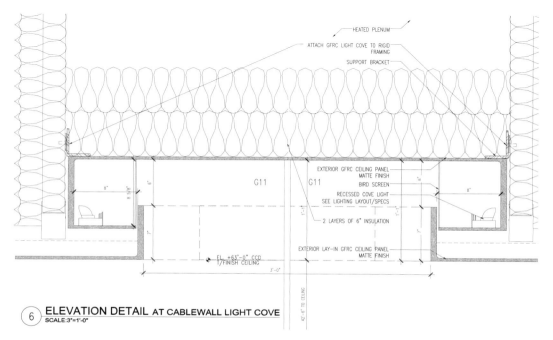

⑥ **ELEVATION DETAIL** AT CABLEWALL LIGHT COVE
SCALE:3"=1'-0"

新办公楼占地约13万平方米，楼内建有崭新的大厅和拱顶走廊，既与街景相互映衬，也使中心大楼的公共入口活力十足。大厅与拱顶走廊咫尺相邻，顶部是网格纹的天花板。在天花板与灯光的映衬下，内侧大厅和外侧拱道似乎天衣无缝地连接了起来。

设计师巧妙地选用LED条状灯管，并将其嵌入天花板网格线凹口中，清晰地勾勒出了屋顶的轮廓。由于这些隐形光源分布于室内外各个角落，所以对质量要求很高，能适应不同的天气。

墙壁凹口内装有双管嵌入式的陶瓷金卤强光灯，使楼内温度和光照十分宜人。将这些灯具隐匿于凹口内，则使天花板看上去整洁如新。

嵌入式金卤强光灯装配有散射透镜，形成

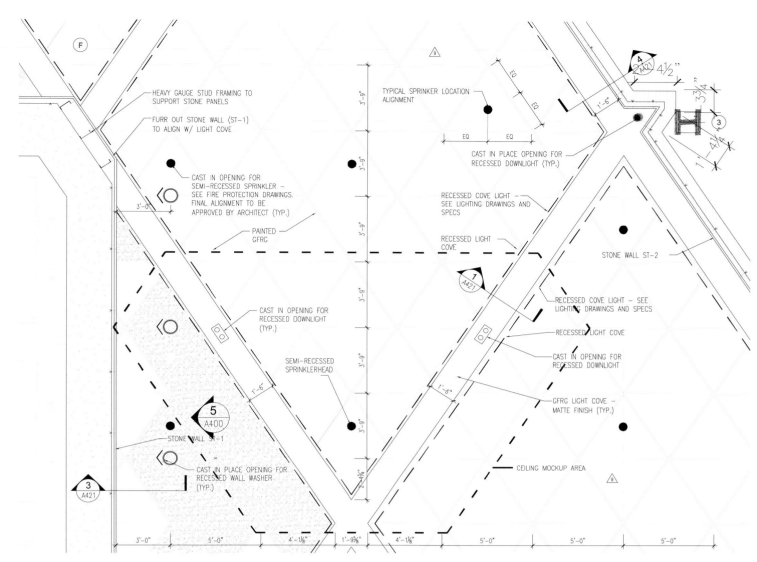

② **PARTIAL REFLECTED CEILING PLAN** AT EAST LOBBY
SCALE:1/2"=1'-0"

一幕柔和的光晕，照亮了周边的砖石。周围的凹口采用条状荧光灯进行间接照明，营造出天花板"悬浮"于侧墙之上的效果。所有灯管均采用目前常用的3 000开色温的。而电梯大厅的照明则依赖于隐匿的陶瓷金卤泛光灯。这些灯具全都内嵌于心墙的凹槽内。间接照明系统使大楼暖意融融、热情洋溢。正是清晰的几何造型和综合性的照明措施成就了这座热情四溢的建筑。

该项目获得了LEED金奖认证。所有光源高性能，低维护。大厅灯光的功率密度达到了12.9瓦/平方米。

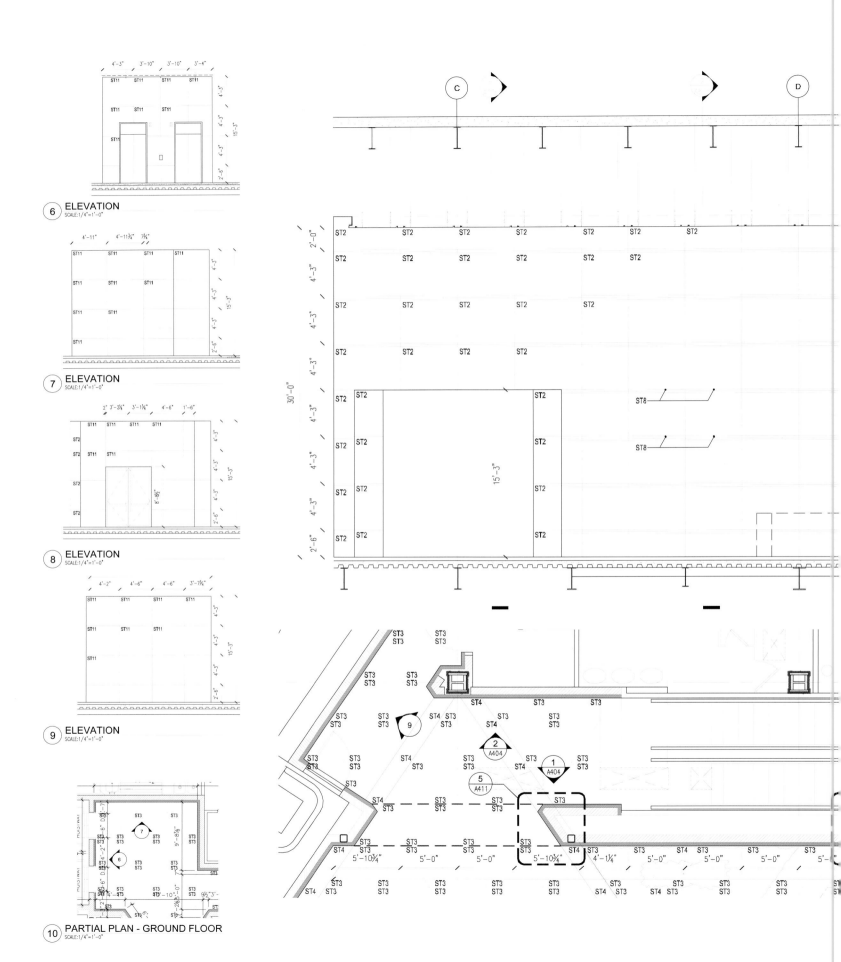

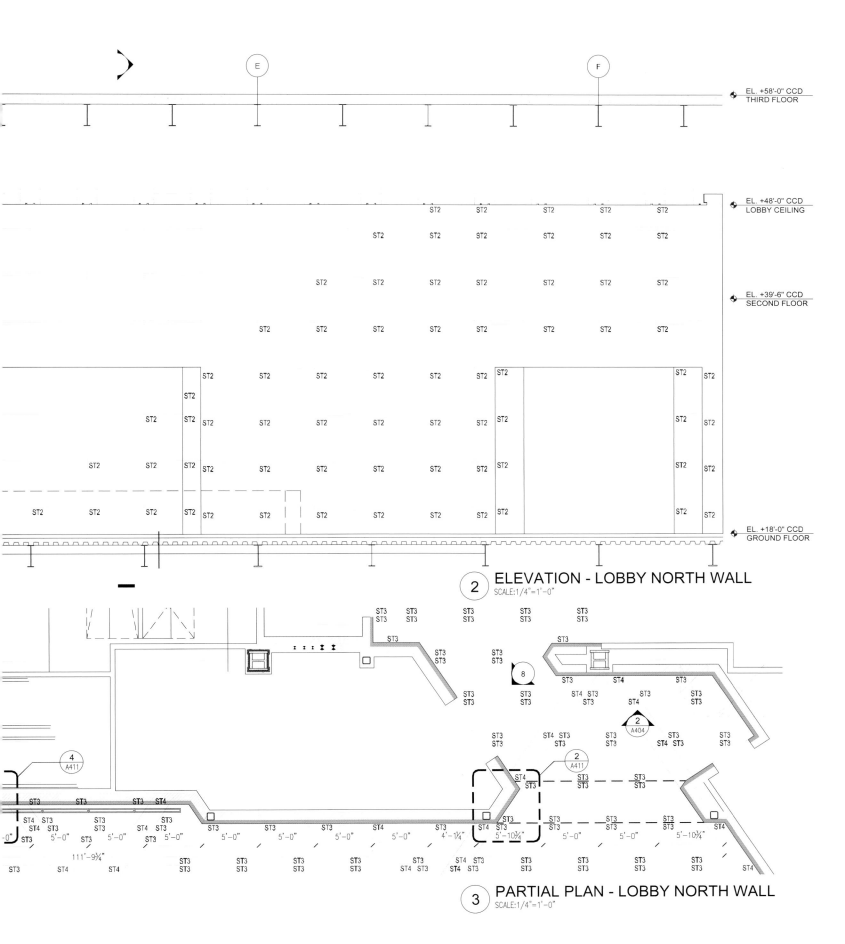

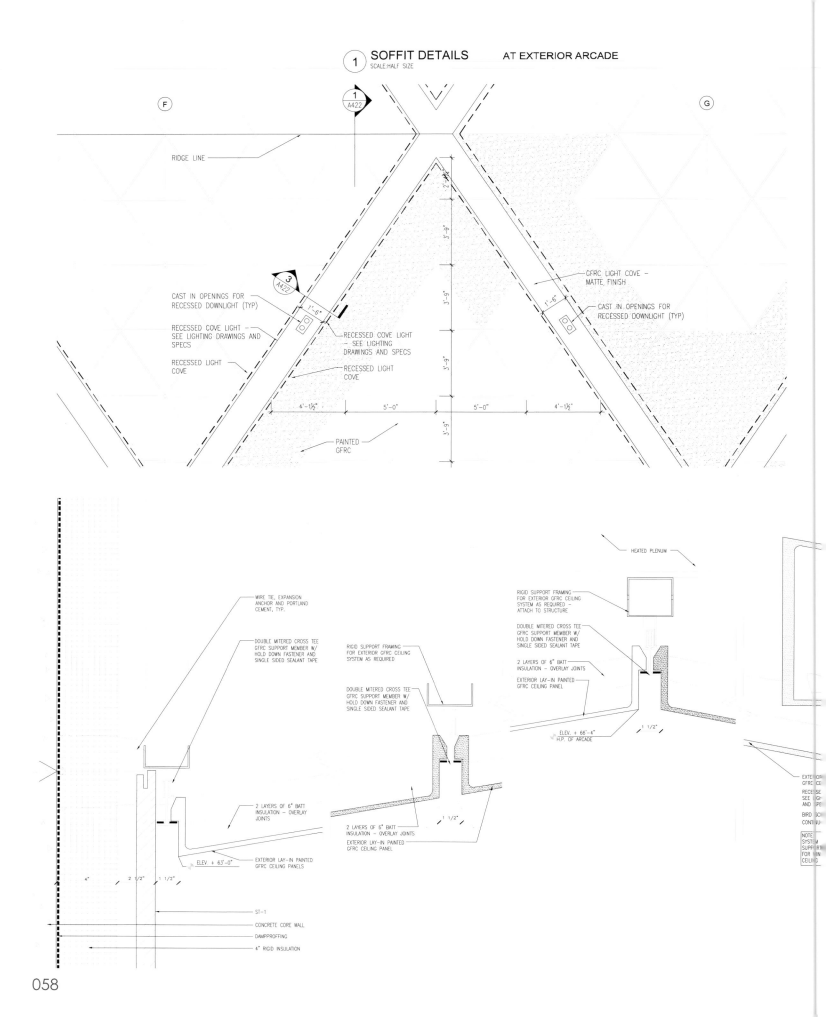

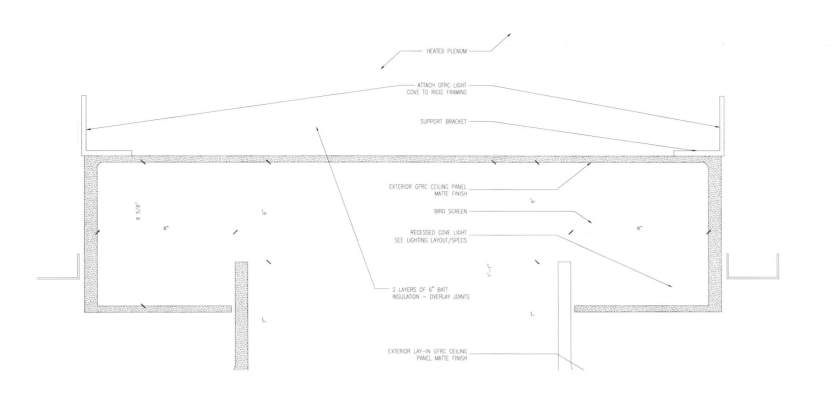
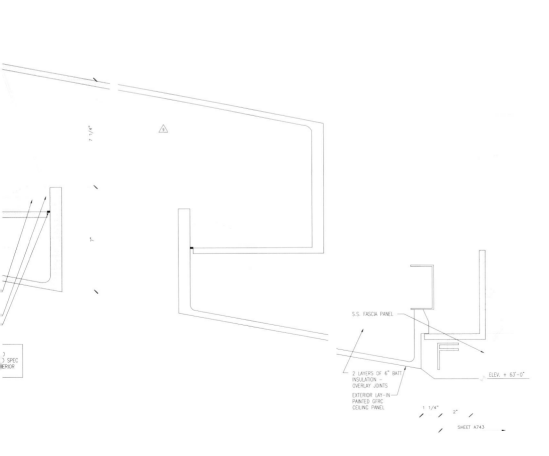

# Locarno: The Ideal City

## 哲学与三色光源
### ——洛迦诺：理想城

**Credits**

Location: Locarno, Swiss
Client: Società Elettrica Sopracenerina SES
Lighting Concept: Prof. Vittorio Storaro - Arch. Francesca Storaro
Lighting Design: Arch. Francesca Storaro (Aild, AIDI, PLDA, IALD)
Luminaires: Erco - Philips - Bega- Simes - iGuzzini
Management of Works: Arch. Antonio Pedrazzini – SES
Projectors: Erco - Philips - Bega
Photography: Massimo Proietti

In 2003, the Swiss electrical power company Società Elettrica Sopracenerina (SES) chose to present the city of Locarno with a project of illuminations to mark the 100th anniversary of its foundation, showcasing the four main symbols of the city by bathing them in light: the historic building housing the headquarters of SES, Castello Visconteo, Piazza Sant'Antonio, and the Madonna del Sasso sanctuary (in progress). Castello Visconteo and SES headquarters (external façade and internal courtyard) were completed in 2004. Piazza Sant'Antonio, including the Church of Sant'Antonio Abate, Casa Rusca and the fountain, have been recently realized, on August 2010.

The lighting concept is based on a parallel between Locarno and Plato's ideal city (ideal = balance).
The search for an ideal form of government is the fundamental focus of PLATO's philosophical analysis. The doctrine of the moral virtues – the disposition of the Soul that steers individual will towards good–has ancient origins and is expressed in Plato's writings through the theory of the TRIPARTITE SOUL. Plato established a strict correspondence between the parts of the soul (concupiscent, irascible and rational) and the social classes that constitute the ideal State (artisans, guardians and rulers). In terms of this network of relationships, the State – just like the soul of man – has its virtues. For each social class, each with its own PLACE, there is a virtue, just as with the soul of man. The virtue of the artisan is Temperance, of the guardian Strength, of the rulers Wisdom; it is only through the balance of these that, according to Plato, Justice can exist.

The virtues can be visualised through representations of symbolic figures. Indeed, there is a clear link

between Man and his VIRTUES, and Light and its COLOURS. It is through this visual metaphor that the Sopracenerina company building – bathing itself in light – will show the whole city, through its Colours, the virtues of the city itself. The product of a fantasy located in an intermediate realm between the ideal and the real, Locarno is turning the spotlights on itself in an ancient/new space through the world of Light and Colour.

The concept of balance inspired the project; given the existence of a relationship between man and his virtues, as much can be visualised by means of light and its colours.

Primary colours (red, green and blue) were used for the moral virtues; the hues of red illuminate the Castello Visconteo, home of the guardians; green is the prevalent colour in piazza S. Antonio, where the

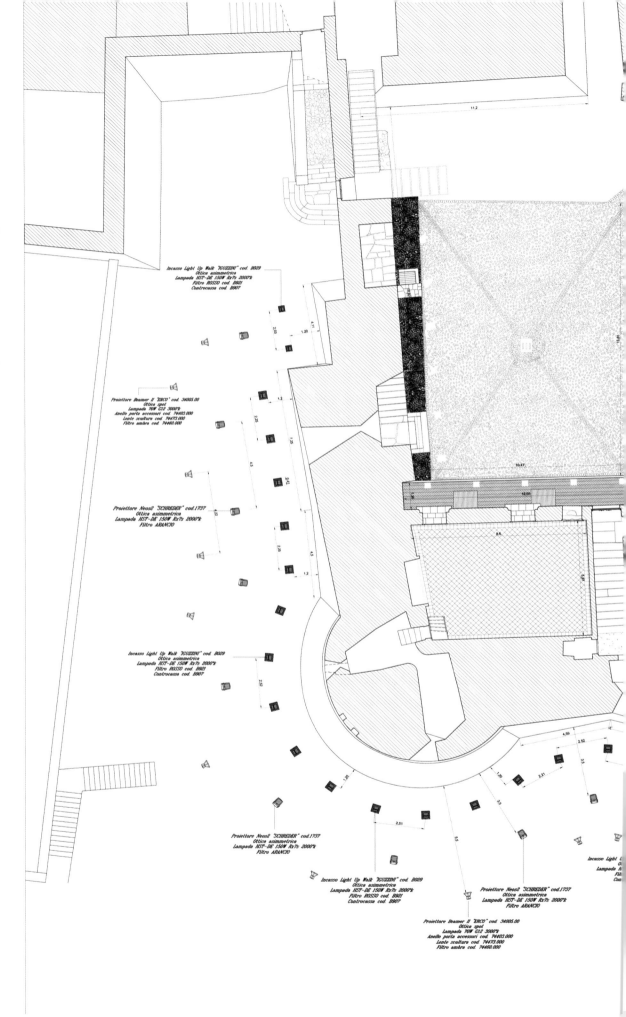

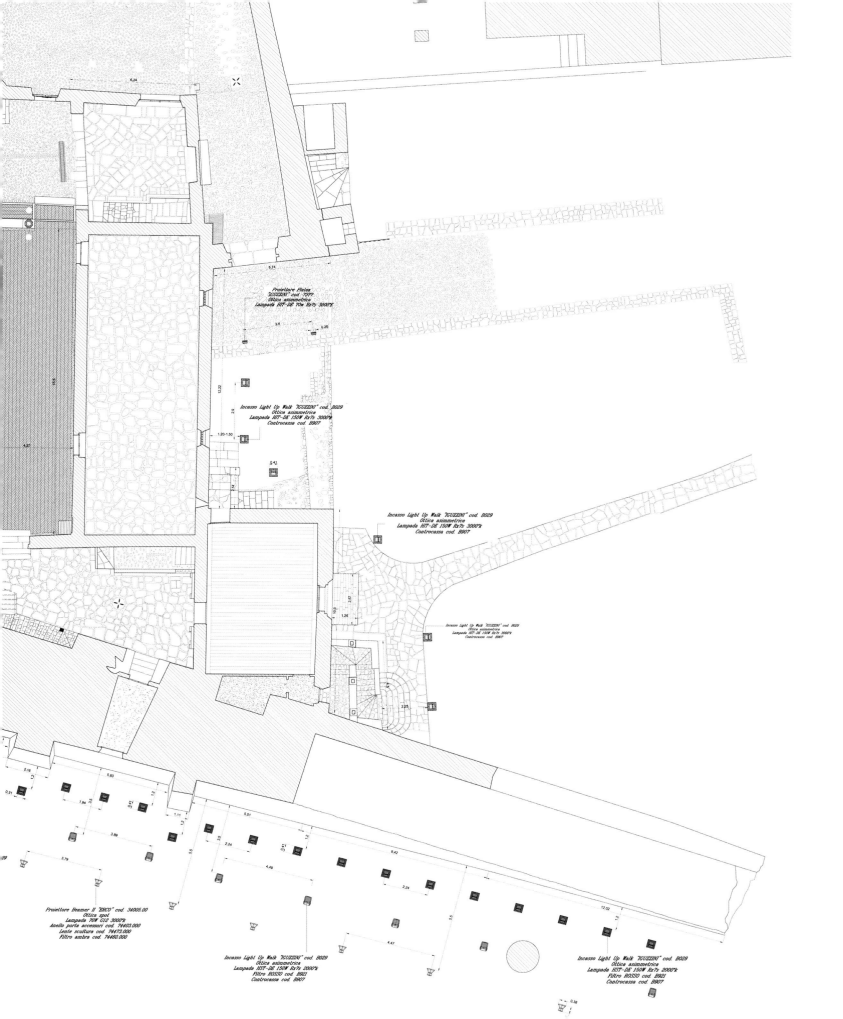

people (artisans and merchants) would congregate; and blue, for the interior of the ancient Government palazzo (today housing the offices of the Società Elettrica Sopracenerina), seat of the wise. The exterior is illuminated by white, the sum of all colours, representing justice.

2003年，瑞士国家电力有限公司（SES）选择给洛迦诺城市基金会的百年庆典安装一套新的照明系统，在灯光的沐浴下突出了城市四个主要标志：SES总部的历史建筑、魏斯康蒂城堡、圣安东尼奥广场和萨索圣寺院（在建中）。圣安东尼奥广场和SES总部（外立面和内部庭院）于2004年完工。圣安东尼奥广场包括圣安东尼奥阿贝特教堂、卡萨鲁斯卡和喷泉，于2010年8月完工。

照明的设计概念使洛迦诺和柏拉图的理想城市（理想＝平衡）并存。政府寻找的完美形式是基于柏拉图的哲学理念。伦理德性的教条——灵魂指引个人通向正义——起源于柏拉图的"三重灵魂"理论。柏拉图要求灵魂的各个部分（贪欲、愤怒和理智）和社会等级要有严格的一致性，以

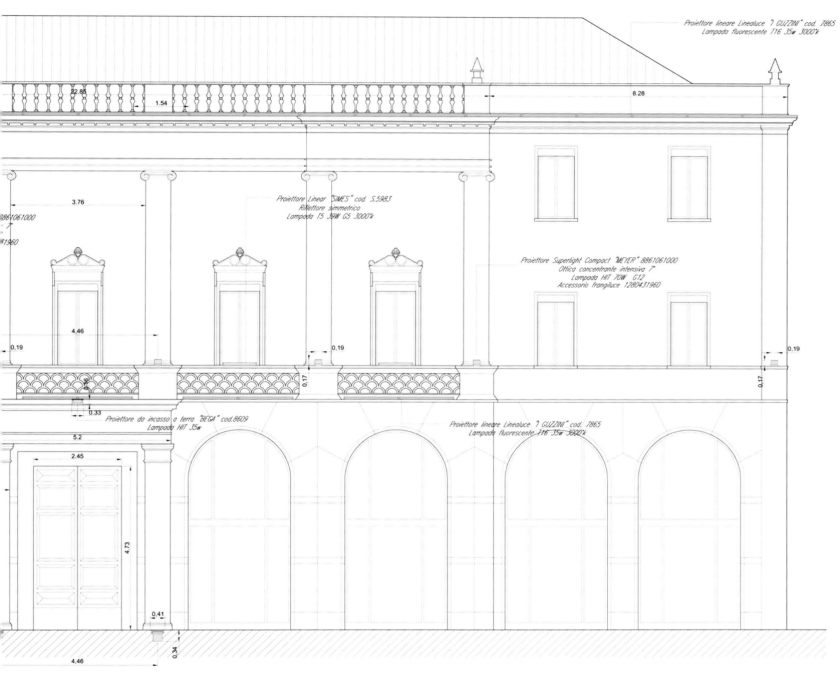

此来构筑一个理想国（由工匠、监护人和统治者组成）。依照这种关系网络，这个国家就像是人类的灵魂，有着各自不同的功能。每个等级都有各自的位置，集合了人类灵魂的优点。

工匠的温和、监护人的权力和统治者的智慧，柏拉图认为只有这三者达到平衡时，正义才能得以实现。

这些德性可以通过象征符号使之形象化。事实上，人类和他的德性就如同灯光和它的颜色之间有着千丝万缕的联系。电力公司大楼通过这种视觉隐喻——沐浴在灯光之中——通过颜色来展示城市本身的德性。这个梦幻的产品构筑了理想和现实的世界。洛迦诺在这个古老/全新的空间中通过灯光与色彩的世界使自身成为焦点。

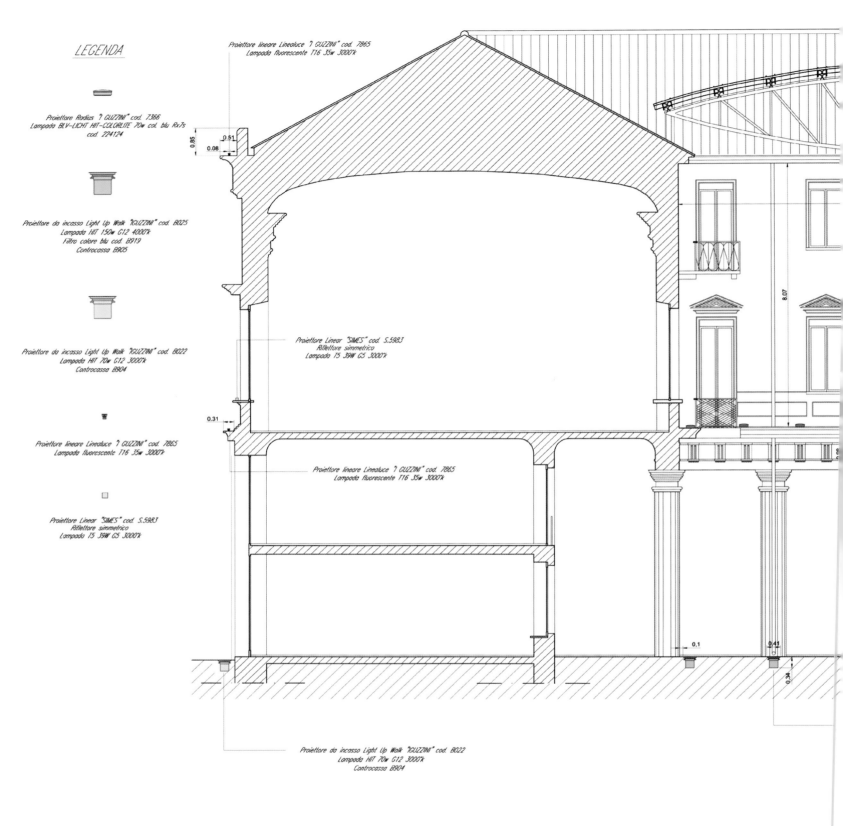

平衡的概念影响了项目的设计，通过灯光和颜色给予人类和他的德性一种可视的关联。

三种基本色（红、绿、蓝）用来描述道德品性——红色照亮了魏斯康蒂城堡（监护人的家园）；绿色用于圣安东尼奥广场，是人民（工匠和商人）聚集的地方；蓝色用于古老的政府大厦（现SES办公楼）内部，代表智慧。外部使用了白色，所有颜色的集合，代表着正义。

Proiettore Radius "7 GUZZINI" cod. 7366
lampada BLV-LICHT HIT-COLORLITE 70w col. blu Px7s

Incasso Light Up Walk "IGUZZINI" cod. B025
lampada HIT 150w G12 4000°k
filtro colore blu cod. B919
Controcassa B905

# Palazzo d'Arnolfo

## 多色调白光的精彩演绎
——阿诺尔夫宫殿

**Credits**
Client: Commune of St. Giovanni Valdarno
Architectural Design: Arnolfo Di Cambio
Lighting Design: Arch. Francesca Storaro
Luminaires: iGuzzini illuminazione Spa

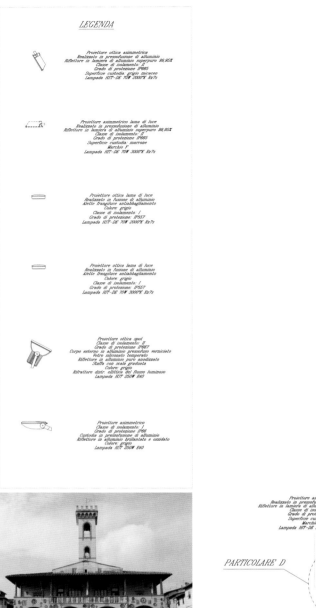

PROSPETTO SU LARGO ARNOLFO DI CAMBIO

The geometric rigour that regulates the town-planning of St. Giovanni, has its centre just in the Palazzo, that is in the main square, in axial position in order to favour the vision, along the centre-line. On the occasion of the exhibition "Arnolfo Urbanista", edited by the Prof. E. Guidoni, and of the Conference: "Medieval New City St. Giovanni Valdarno, Tuscany and Europe" organized by the Commune of St. Giovanni Valdarno, and on the occasion of the Seventh Centenary of the death of Arnolfo di Cambio, there has been the proposal to realize a project of permanent lighting of the Palazzo D'Arnolfo.
The use of a white light of various tonalities will give prominence to the different parts of the Palazzo: those realized by Arnolfo, will be lightened by a white light, characterized by a warm tonality (solar); a white light, with a cold tonality (lunar), will be used to lighten the following interventions of the arcades. Also the tower, the element of completion of the Palazzo, that was already present in the original sketch, will be lightened by a warm light in the tonalities of the orange.
The portico, a void realized soon after, today represents an important element of the city life, and

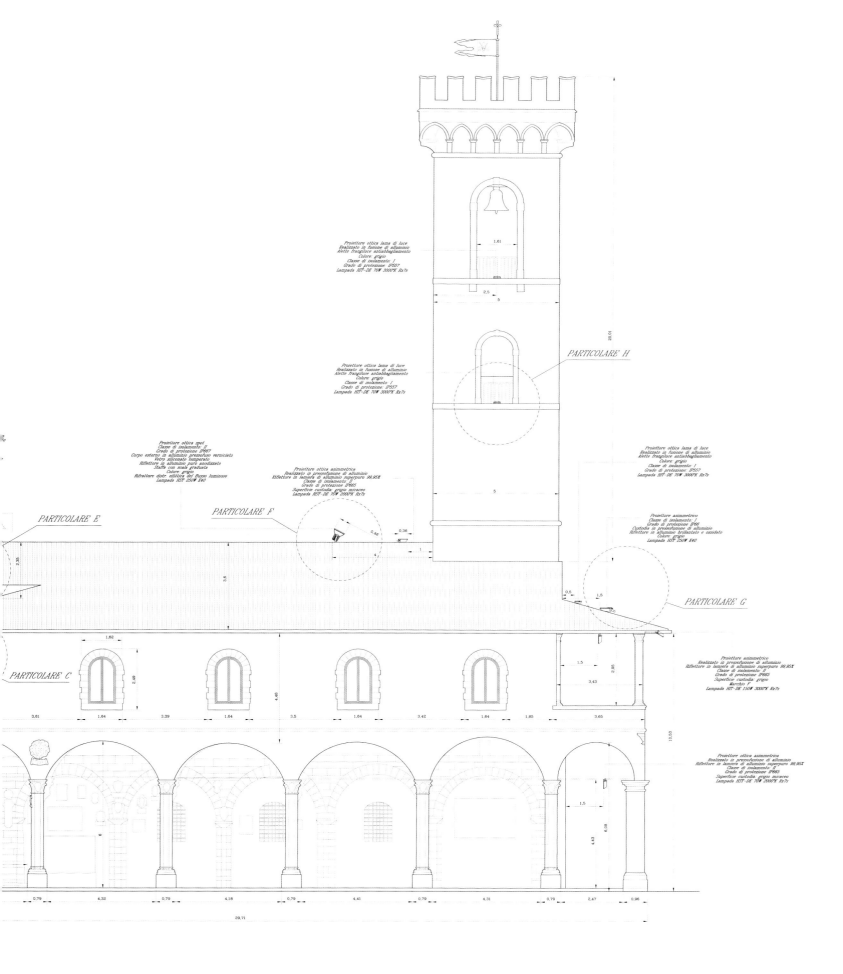

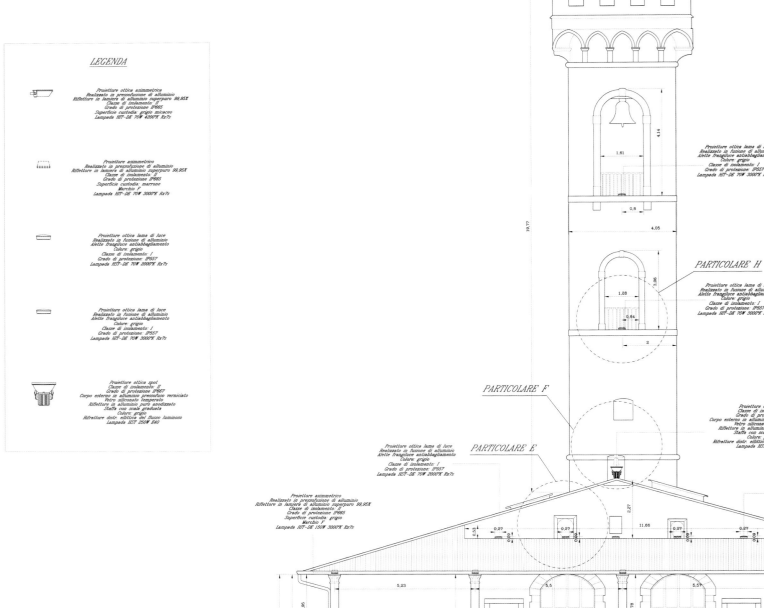

it will be lightened by a cold light, blue.
The overhanging façade will be lightened instead, by a neutral white light as element of union of the whole architecture.
The side facades will be left in penumbra to underline the axle of reference that is constituted by the façade on Piazza Cavour and on Piazza Masaccio.
So differentiating the nighttime vision from the diurnal one, it is possible to underline the original architecture of the project of Arnolfo that has brought again to light.

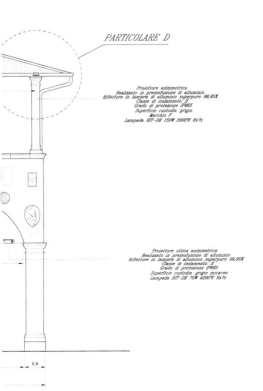

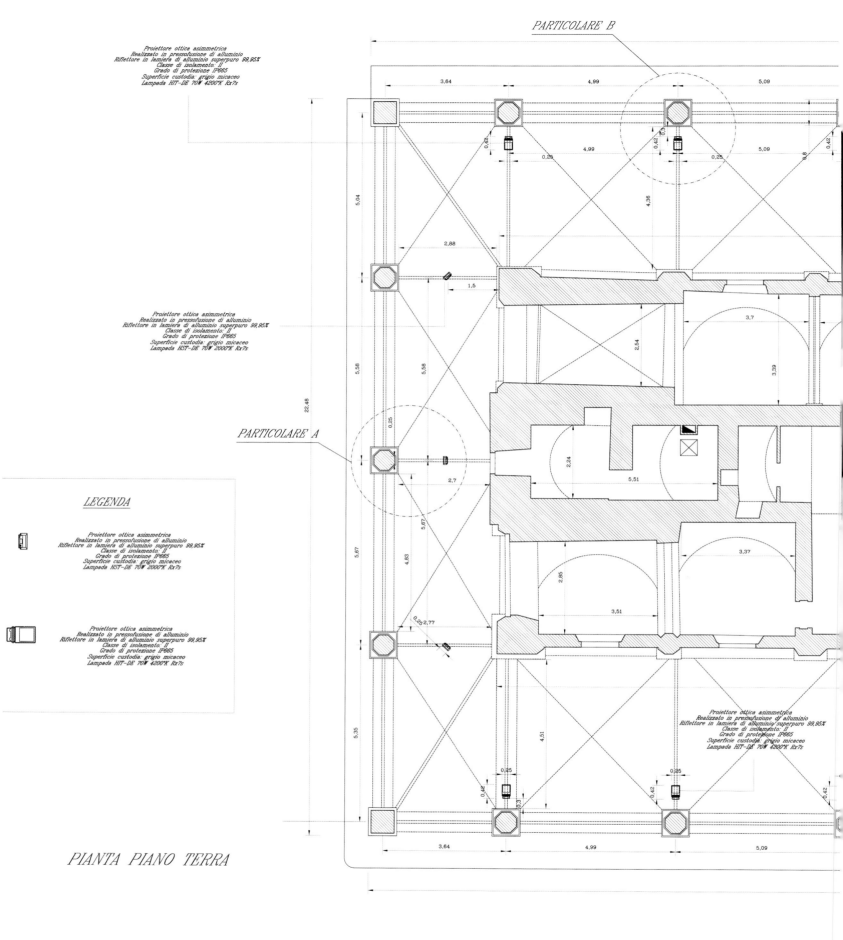

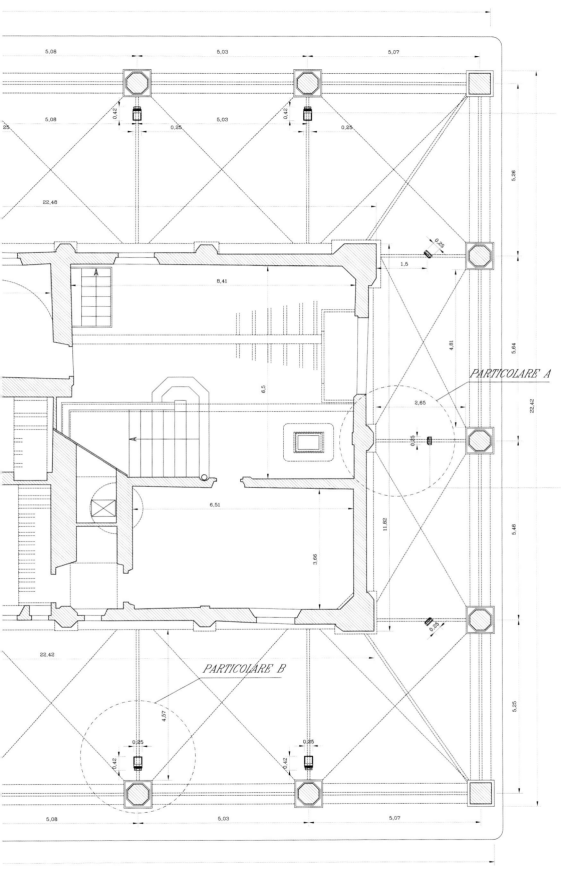

为了使城市更具观赏性，圣乔瓦尼市的城市规划严格按照中心对称要求设计。它的中心便是阿诺尔夫宫殿及其主广场。

在由圭多尼博士策划的"阿诺尔夫城市规划专家"展览、圣乔瓦尼瓦尔达诺公社举办的"中世纪新城托斯卡纳，圣乔瓦尼瓦尔达诺和欧洲"会议行将举行、艺术家阿诺尔夫·迪·坎比奥诞辰700周年之际，阿诺尔夫宫殿启动了一个永久性的照明项目。

使用多色调的白光强化了大楼内不同区域的差异性：由阿诺尔夫设计建造的部分是暖色调的白光；后面的拱廊则使用了冷色调的白光。

作为完成宫殿的最后元素，塔楼将使用暖色调的橙色光。

后来才被挖出的柱廊如今成为了城市生活密不可分的部分，使用了冰冷的蓝光。

悬空的立面使用了中性的白光，作为整个建筑的整合元素。

侧面留下了加富尔广场和马萨乔广场的半影，强调了建筑的轴线作用。

白天和夜晚的景观是如此的不同，灯光的设计将阿诺尔夫的建筑精髓再次展现出来。

# Old DC Courthouse

## 照明之于国家地标的修复
——华盛顿哥伦比亚特区法院

**Credits**
Architects: Domingo Gonzalez Associates – Domingo Gonzalez, AC Hickox, and Carolyn Schultz
Awards:
2010 IIDA/IES Lighting Award of Merit
2010 Washington Building Congress Award, Lighting
2010 US General Services Administration (GSA) Design Award
2010 AIA Justice Facilities Review Award
2010 Associated General Contractors of America National Merit
2010 Structural Engineers Association Metro. DC Project of the Year
2009 DC Award of Excellence, Historic Preservation
2009 DC AIA Award of Merit Historic Resources
2009 Mid-Atlantic Construction Award for Government Projects
2009 McGraw-Hill Construction's Best of the Best Award

Maintaining historic fidelity while meeting modern criteria are but two challenges encountered when faced with the historic preservation of a landmarked courthouse and its adaptive re-use as a federal courthouse facility – especially one where President Lincoln himself was known to speak from the building's balcony. Mix in the need to maintain a low-energy profile, and the expectation to retain historic luminaires; add budget and time constraints; and a complex, yet rewarding, project can emerge.
A six-year collaboration between Domingo Gonzalez Associates (DGA) and Beyer Blinder Belle Architects

culminated in 2009 with the completed $99 million historic preservation adaptive reuse effort of the landmark Old DC City Hall (after a decade of vacancy) for use as a modern courthouse facility for the DC Court of Appeals. The lighting design was required to meet the illumination and uniformity requirements of the US GSA Public Buildings Standards, as well as the energy use constraints inherent in ASHRAE 90.1 for modern courthouses – while maintaining fidelity to the building's significant legacy.

Our scope of work included the development of lighting design strategies for public areas, courtrooms, judges chambers, library, atrium rotunda and exterior site lighting. The project was also designed to meet requirements for LEED Certification (Leadership in Energy and Environmental Design) by the US Green Building Council.

Our services included a programming analysis of

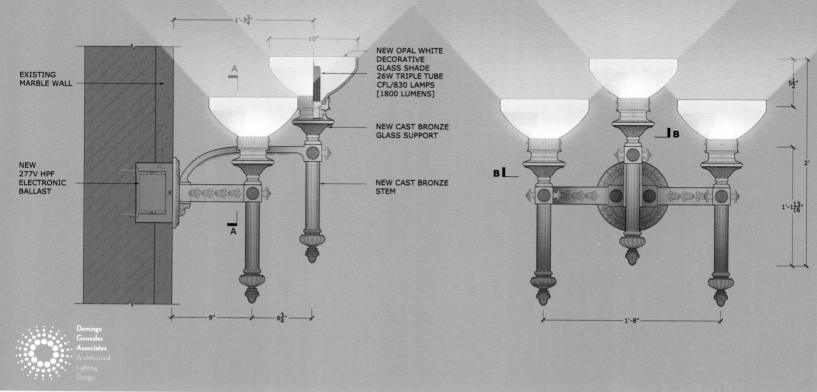
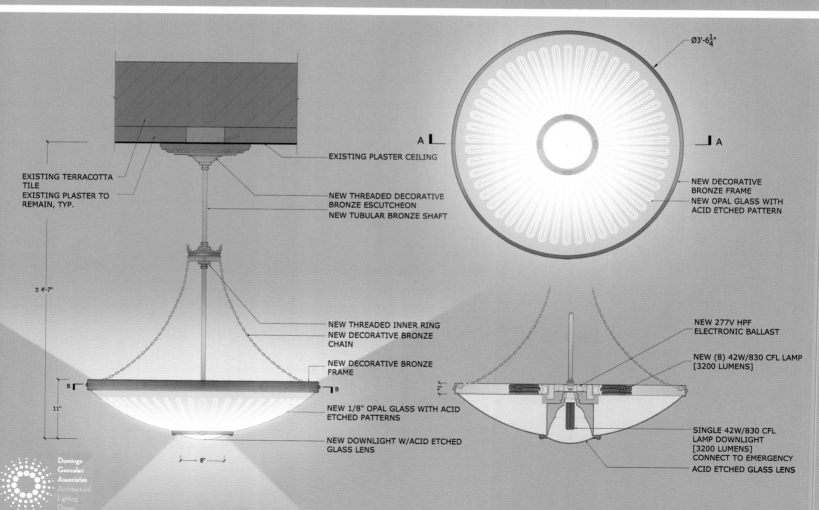

light level requirements, cost control, and detailed computer-generated lighting calculations, as well as provision of specifications for lighting equipment and control systems. Perhaps the most rewarding aspect of the project was the opportunity to develop custom replicas of historic luminaires. The original terracotta ceilings in the historic courtrooms offered few mounting locations. To meet stringent US Courts requirements, DGA designed chandeliers to evoke a late-19th century sensibility. A prequalified list of manufacturers competed to fabricate the luminaires. Elsewhere, existing luminaires were given new life, refurbished to accommodate new energy-efficient light sources.

维护历史的真实性同时又要符合现代标准，成为法院这一历史性地标建筑的保护及其修复与重新用于联邦法院设施所面临的两大挑战，尤其是以林肯总统在这一建筑的阳台上发表演说而为世人知晓的历史建筑。维护低能耗设备，保留历史性灯具，增加预算，突破并拉长工期：由此诞生了一个复杂却富于回报的项目。

DGA照明设计公司和BBB建筑公司之间六年的合作于2009年达到顶峰，实现了投资额为9 900万美元的华盛顿哥伦比亚特区法院大楼(空置了十年之后)的历史性地标建筑保护修复与重新用于华盛顿上诉法院的现代法院配套设施。照明设计必须符合美国总务署(GSA)公共建筑标准关于照明和统一的规定，同时遵循美国采暖、制冷与空调工程师学会(ASHRAE)标准90.1的能量效率要求，不仅体现现代感，还要维护这一重要建筑遗产的真实性。

工程范围包括各区域的照明设计策略，如公共区域、法庭、法官室、图书馆、中庭、圆形中厅及室外场地。项目设计还力图符合美国绿色建筑协会的LEED认证标准(领先能源与环境设计的评级系统)。

服务范围包括光级要求和成本控制的程序分析、电脑输出细部照明计算、照明设备和控制系统的

技术规格制定。或许，工程最具回报的方面在于有机会研发历史性灯具的特殊复制品。原历史法庭的赤陶天花板几乎没有安装灯具的位置。为了符合美国法院的紧急要求，DGA照明设计公司设计了枝形吊灯以唤起人们对19世纪后期的感受。经资格预审的制造商对灯具的制造进行竞标。此外，对原灯具进行翻新，装置了新型节能灯，使其焕然一新。

# "Lightstream" Dallas Convention Center

## "光流"
## ——达拉斯会展中心

**Credits**
Location: Dallas, Texas, USA
Lobby Sculpture: 800' x 60' x 20', Dichroic Glass, Stainless Steel Cables & Hardware, Aluminum Extrusions, Computerized Lighting
Client: City and County of Dallas, Texas
Lighting Consultant: Craig Marquardt
Lighting Control System: ETC/Barbizon
Project Manager/Design Collaborator: John Rogers
Project Assistant: Oanh Tran
Metal Fabrication: Albina Pipe Bending, Portland, OR
Glass Fabrication: Haefker O'Neill Studio, Portland, OR
Architect: S.O.M., Chicago / H.K.S., Dallas
Photography: Ed Carpenter, John Rogers

A river of light glows overhead. Streams of color pulse in waves of illumination. Floating lines in space suggest surface and depth. Tinted skies roll with swells of luminescence. "Lightstream" is an 800' long kinetic installation in the lobby of Dallas' new convention center addition.

This piece builds on the power of an extremely long space by creating continuous kinetic light effects along the entire ceiling of the Convention Center expansion lobbies. By placing scores of floating dichroic "light sticks" suspended in an apparently random arrangement, flowing pools of color are painted across the ceiling, and a feeling of continuous change and process is created. Like sticks awash in a stream of light, they create a layer of secondary texture. Their randomness is juxtaposed to the order of the architectural grid, downlights, and suspension elements. Order underlies chaos. Crowds of conventioneers form an additional random element, further layering the composition. Microprocessors control the movement and fading of the light sequences in subtle watery patterns, with waves of light washing from one end to the other, or sometimes crossing in the middle, rippling like liquid in a wave tank. Like the sound of moving water, the play of light is refreshing, energizing.

光流淌在空中,色彩的溪流涌动在照明波浪里。空间里漂浮的线条框出了表面和深度。着色的天空席卷了凸起的发光体。"光流"这一富于动态美的装置长约244米,安装在达拉斯的新会展中心入口大厅。

这件小品构建在一个特别狭长的空间里,沿着会展中心扩建入口大厅的整面天花板,营造出连续而具动态感的灯光效果。大量漂浮的二向色"灯棍"做悬挂状,呈现出显而易见的随意布局,流动的色彩池喷绘于整面天花板,营造出连续变化及其过程演变的感觉。棍子仿佛被灯流冲刷,创造出一层次级纹理。它们的随意性与建筑方格网、向下射的灯、悬承元素的次序感并置。次序突显混乱。与会者人群走过这个随意元素构成的附属小品,更是增添了几分混乱感。

微处理器控制着光指令序列,使光的运动和褪变呈现静谧的水型,光的波浪从一端荡漾至另一端,有时从中央横过,泛起的涟漪仿佛水滴落入波浪池里。宛若水体流动的声音,光的灵动令人焕然一新,充满活力。

# Roca Barcelona Gallery

黄昏里的"面纱"
——乐家巴塞罗那展厅

**Credits**
Location: Barcelona, Spain
Scale: 750 m²
Office: artec3 Lighting Design
Watts per Square Meter: 21 W/m²
Architects: OAB (BorjaFerrater, LucíaFerrater, Carlos Ferrater)
Software: Visionarte (Intesis)
Designers on Project: MauriciGinés, Jesus Gonzalez, Cristina Salicio, IvanEscutia

The lighting system for the new Roca Barcelona Gallery was designed to transmit the importance of water in our civilization, the value and comfort that has provided in our homes throughout the history. Using light we mimic water behaviors to create a relaxed atmosphere for the contemplation of the bathroom product and spaces in the exhibition. We wanted to transmit ideas of intimacy and hygiene.

The use of artificial light on the glass façade at dusk acts as a veil, and creates a sensations of fluidity, tranquility, smooth dynamism, an almost hipnotic mood, to both the exterior and interior spaces. As a result, the gallery visitor (as well as the passerby) enters in a more comfortable atmosphere that provides him a better building experience.

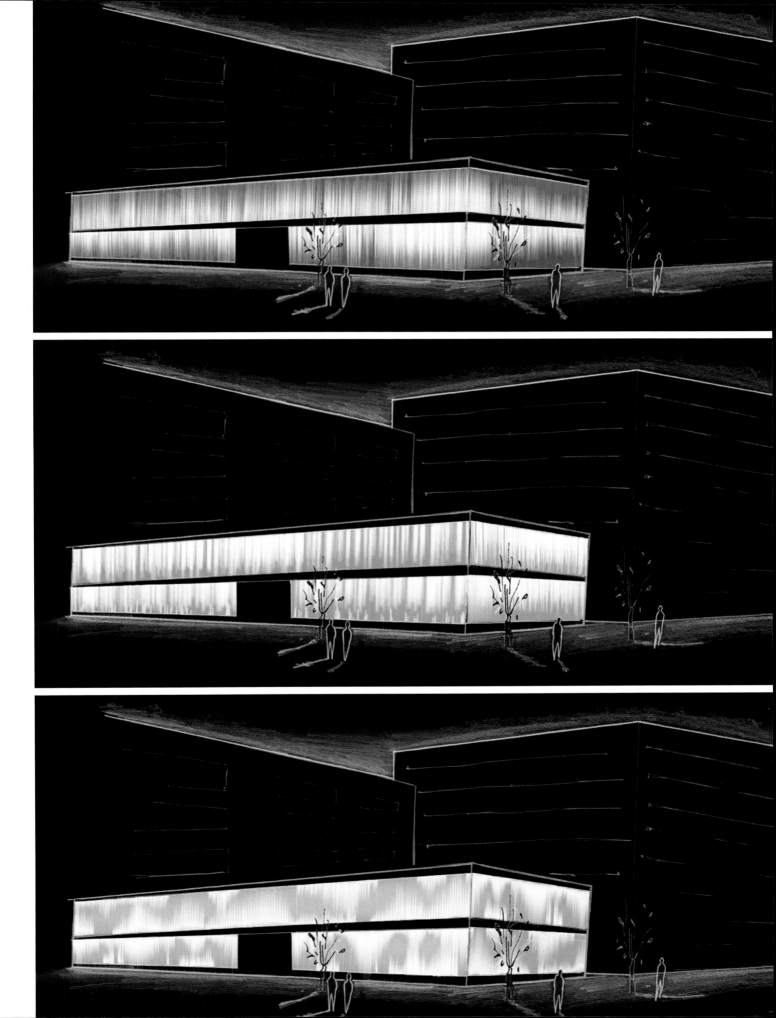

新建的乐家巴塞罗那展厅的照明设计，旨在突显人类文明进程中水元素的重要性，以及整个历史进程里家庭的价值和舒适感。

灯光设计模拟了水的运动，在展厅的卫浴产品空间里，营造出富于冥想的轻松氛围。设计师意在向观者传递亲密和卫生的理念。

玻璃幕墙的人造灯饰，在黄昏里俨然一款面纱，流动、静谧、畅快的活力，这种催眠的状态弥漫于整个室内和室外空间。展厅游客(包括路人) 只要一接近或步入展厅，就体验到更为舒适的建筑氛围。

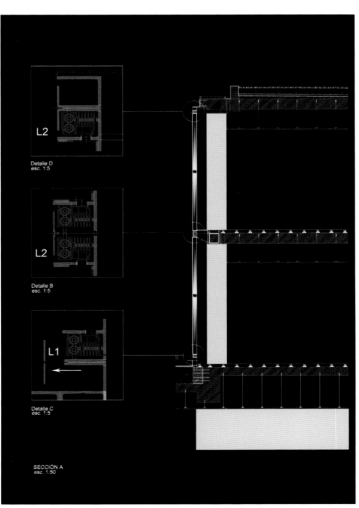

SECCIÓN A
esc. 1:50

Detalle D
esc. 1:5

Detalle B
esc. 1:5

Detalle C
esc. 1:5

# Pio Palace in Carpi

与建筑相融合
——卡尔皮皮奥宫殿

**Credits**
Lighting Design: Barbara Balestreri Lighting Design
Photography: Balestreri Studio

Pio Palace in Carpi is an example where both the architectural space and the fit-out have close relationship with the lighting system proposed.
The principal object of our design was to include elements that were not invasive in terms of aesthetics and conservation of the rooms, with frescos completely covering both walls and ceiling.
We have developed a new custom system that could meet different environmental demands.
"Mosaico" is been designed for this purpose (product made by Viabizzuno company):
unit from the ground, with a wall mounting plate, which connects the pedestal to electrical outlets.
The highly flexible system incorporates an accent light with adjustable halogen spotlights, a functional diffuse light provide by fluorescent lamps that highlights the precious decoration of the walls and timber ceiling.
Also the open gallery on the main floor, with an architectural rhythm articulated with vaults and window represented an opportunity for the construction of a new lighting fixture.
While on the one hand it was necessary to add a formal element in harmony with the architecture, on the other it was important to make this special fitting to revive the concept of suspended lighting, typically used for lighting galleries but in a new formal guise to give emphasis to the environment in which it was installed.
The result is "Principio" (product made by Viabizzuno company): light fixture suspended.
The fittings are suspended from the centre of each of the gallery's vault, creating a captivating visual rhythm and at the same time providing adequate warm and welcoming lighting that can both enhance the architecture of the space and make it more usable.
The fitting has highly original appearance and incorporates the option of producing two different types of lighting an atmospheric light and a more functional and direct, downward light, for use also in conjunction with temporary exhibition.

意大利卡尔皮皮奥宫殿的照明系统与建筑空间、装修有着紧密的关系。

设计的主旨在于，所融入的元素不能体现出侵略性，如美学、房间的保护、墙面和天花板的整体壁画等。

设计师还设计了一款新型定制系统，用于满足不同的环境需求。

基于以上目标设计了"马赛克画"（产品为Viabizzuno公司制造）：壁画从地面向上，由连接基座和电源插座的壁上安装板支撑。

高度灵活的系统整合了可调卤素聚光灯和功能性漫射光来营造强化的效果，后者由荧光灯发出，聚焦了墙面和木质天花板的珍贵装潢。

主楼层的开放式画廊，以拱顶和窗户为建筑旋律，为新照明灯具创造了安装机会。

一方面，增加一个与建筑相融合的规则元素十分有必要；另一方面，利用这个特殊装置复兴悬挂灯的理念也十分重要，特别是画廊照明，而不是一种强调安装环境的新型规则掩饰。

最后引入了"原理"（产品为Viabizzuno公司制造）：悬挂灯具。

灯具悬挂于画廊每个拱顶的中央，营造了一种迷人的视觉旋律，同时提供了足够的暖度和亲切的照明，不仅美化了建筑空间，还增加其使用功能。

灯具的外形独具原创性，产生两种不同的照明，一种用于营造氛围，另一种更具功能性，其直接的、向下投射的光能够在临时展览里发挥作用。

# Klimahaus 8° Ost, Bremerhaven

## LED点阵灯的演绎
## ——不来梅东经八度气候馆

**Credits**

Location: Bremerhaven, Germany
Scale: 7,700 m² (Facade)
Client: Bremerhavener Entwicklungsgesellschaft; Alter/Neuer Hafen mbH & Co KG, Bremerhaven
Lighting Design Facade / Exterior: pfarré lighting design, Munich; Gerd Pfarré, Katrin Rohr, Katja Möbs
Architects: Klumpp Architekten, Bremen, agn Niederberghaus & Partner GmbH, Ibbenbüren; Wolf Bartuszat
Landscape Architects: Latz & Partner, Kranzberg/London; Tilman Latz, Oliver Keil
LED Custom Products: Feno GmbH, Munich
Leading Idea Klimahaus: Petri + Thiemann GmbH, Bremen
Energy Concept: Transsolar Energietechnik GmbH, Stuttgart
Photographer: Markus Tollhopf, Hamburg

The Klimahaus 8° Ost is situated on the longitude 8° East. Based on a mix of science, entertainment, and edutainment, it represents selected climate zones along this longitude, around our globe. With a length of 125 metres and with its totally glazed facade, the building is a key landmark of the city of Bremerhaven. The facade lighting concept was composed with 1,900 individually addressed, cold-white LED. At night, the LEDs enhance the form and content of the project. Various scenes, developed according to the Klimahaus concept, have been created to reflect the themes of this ambitious and successful project.

### Facade Lighting Design

The Klimahaus 8° Ost offers a mix of science and education, and belongs to the "Havenwelten", a large revitalization project in the former port area of the city of Bremerhaven. It invites visitors to undertake a journey through selected climate zones along the

degree 8° East, enhancing the awareness for the global climate, it's features, impacts and problems involved.

pfarré lighting design realized several lighting projects in the Old/New Harbor project within the Havenwelten, so in 2005, they had been invited to present a proposal for the "night time identity" of the Klimahaus, together with 3 other lighting design teams. Pfarré's idea was the only one that faded out the inner volumes of the building. The brief was to create an appropiate night time outdoor image. The design is focussed to the building skin; the idea of a mesh of meridian lines (the 8° East theme), and the use of LED points at the meridians crossings was key part of our presentation. Furthermore, the designers did not only want to create an unrelated media facade, they wanted to bring something from the inside, from the conceptual content, to the outside, to the public, visible at the facade.

Use the free building form as "canvas", the pattern of the glass sheets to create a mesh of single lighting points that can be set into motion by being adressed individually via DMX. Only white light, no color spectacle that tires after a while. Not too much light,

but follow the entire volume with its total length and height. Underline the shape of the project. Fade out all interior structures, stay on the surface.

Being part of this ambitious project, that wants to educate a wider public about our climate conditions and the climate change, pld wanted to use a minimum of energy, of course. With 0.33 W/sqm the designers achieved a complete picture, spanning over 7,700 sqm of building skin. They stick to the budget, which was defined in the preliminary design phase.

pfarré lighting design is convinced about the achievement of a long lasting lighting design. It is not trendy, and it is calm, inviting, almost discreet, but still spectacular. It came out exactly the way the designers intended, in their first presentation. As a "side" effect, apart from the night time image, the building, this gigantic blob, illuminates its surrounding with a total power of 2,600 watts only! However, the maxium wattage, provided by the client, was much higher. Sequences are switched on only for special occasions. This site is not Times Square, and in order to keep the moving lights that play with elements of nature, the oceans, rain, wind and stars in the sky, exclusive, pld agreed with the client to switch it on for special events only. The regular, all-night image is static, but it glows like a huge sponge or sea urchin, carefully embedded in the Havenwelten project, not far from the North Sea.

东经八度气候馆坐落在东经8度。气候馆以科学、娱乐、寓教于乐为出发点,向游客展示了整个地球东经8度的各种气候区。气候馆长125米,外立面均为玻璃铺装,成为不来梅市的一个重要地标建筑。外立面的照明设计采用了1 900个冷白LED灯具。夜间,LED美化了建筑外形,丰富了气候馆的内容。同时,设计师基于气候馆的理念,创造了呼应主题的各种场景。

**外立面照明设计**

东经八度气候馆具有科学和教育意义,是不来梅哈芬市大型旧港口复兴工程"哈芬世界"的一部分。游客在气候馆可以体验一场沿着东经8度的气候区之旅,增进对全球气候,以及各气候区的特征、影响和相关问题的认知。

普弗拉照明设计公司完成了"新/旧港口"工程里的多个照明工程,因此,2005年,他们受邀对气候馆的"夜景形象"工程提出设计方案,同时参与竞标的还有其他三个照明设计团队。普弗拉设计团队是唯一一个提出弱化建筑内部体量的构想的团队。简而言之便是营造一个相得益彰的室外夜景形象。

设计以建筑表皮为重点;经线网格的构思以及经线交叉点上LED点阵灯的使用成为设计陈述的关键部分。同时,设计师并不只是试图营造一个毫无关联的介质外立面,他们希望外立面能够向公众呈现一些室内的理念,使其在外立面上可视。以自由的建筑风格为"画布",结合玻璃面板的样式来营造一个点阵灯的网格,并通过DMX(数字多路复用协议)将其激活。只呈现白色,没有片刻后令人疲倦的色彩。没有太多的光源,只是跟随着整个建筑体量的标高和长度,突显建筑的外形,弱化内部所有结构,仅停留在表面。

作为要向更多公众传授气候条件和气候变化的知识这一雄心万丈的项目的一部分,普弗拉照明设计公司(pld)理所当然考虑将能耗最小化。设计师以每平方米0.33瓦的功率实现了完整的图景,延伸至7 700平方米的建筑表皮。设计师坚持将预算控制在设计初期所制定的范围内。

普弗拉照明设计公司深信照明设计的持久性:无需追赶潮流,它必须是冷静的、亲和的、几近隐匿的,但却仍需具备引人注目的特质。最后的完工效果完全符合设计师的构思。除了塑造了建筑的夜景形象,这个巨大的粒子还照亮了周围环境,其总耗电量仅为2 600瓦!极大地低于业主所提出的最大耗能。

灯光的动画序列只在特殊盛会开启。这个场所并不是纽约时代广场,pld与业主达成一致决定只在特殊盛会的时候才开启动态灯效,使其与各种自然元素相呼应,如海洋、雨、风、天空里的星星等。平日里,建筑在整个夜间的形象都是静止不动的,但却熠熠发光,犹如一块巨型海绵或海胆小心翼翼地嵌入"哈芬世界"这个工程里,与北海咫尺相隔。

# Fondazione Arnaldo Pomodoro, Milan

光与艺术品的对话
——米兰阿纳尔多·波莫多罗基金会展览馆

**Credits**
Location: Milan, Italy
Lighting Design: Barbara Balestreri Lighting Design

The exhibition space of Pomodoro Foundation in Milan is unique because it allows an overall view, that a traditional museum, generally composed of a sequence of environments, it can not allow.

This atypical is because the Pomodoro Foundation was established in a turbine factory, Riva Calzoni. The light in the Foundation becomes a characteristic element because the building envelope has large windows which bring a significant amount of natural light. Artificial light must therefore be worthy alternative and has to reveal aspects that, the natural light, does not show particularly.

The accent light creates a light scenography which can bring out the artworks, the pointing of the beams shows the artwork on the walls and simultaneously give light on the walls. In this way so as to bind and exhibit the same structure but time distinguishing them.

The container and contents communicate each other in different ways during the day; the magic of the natural light work envelop the space and artwork unitary, at night the artificial light enhances the artworks and architecture, so that it becomes itself a true work.

The projectors are extremely versatile on which you can install different kind of accessories like colour filters shapers, anti-glare screens, etc.

The sources, both metal halide and halogen lamps

of different wattages and radius, allow us to mix the light is able to create light scenes featuring different colour temperatures and to introduce greater flexibility to respond to the needs of each artwork, which could be a installations, sculptures, paintings, photographs, etc.

The exhibition capacity of the space and the different characteristics of the exhibits, which may include works from many different dimensions and characteristics, led to use dimmable products.

### Light and Sculpture

The combination of light and sculpture is expressed in the exhibition "Arnaldo Pomodoro. Great works 1972 - 2008", where the incredible variety of the exposed sculptures suggested us to take a different path than the tendency of contemporary art museums where there is diffused light that creates an aseptic space where artworks don't emerge.

The building envelope has been considered as a very importance element,

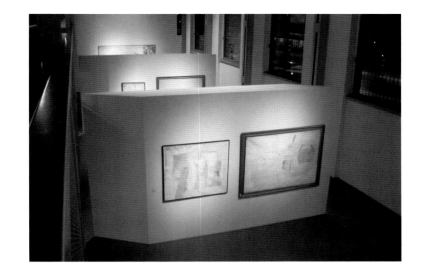

illuminated with large beams of cold light, creating with background a bright contrast that can bring out the sculptures and create perceptions of different light to facilitate the overview of the exhibit design.

Each "artwork" emerges through an accent light, using halogen lamps with spot light, that wrap 360 all the sculptures.

### Light and Installation

A very different approach was adopted for the exhibition "Jannis. Single Act" where the relationship between light and installation was dictated by the artist himself. Kounellis showed us the effect of light more appropriate for its art installation. The artist throughout the period of installation took possession of the space of the Foundation by living day and night. Its aim was to prevail his work on the housing, the Pomodoro Foundation.

The fantastic result was dictated by the choices made by Kounellis for the light effects. Our job was to interpret and apply the thought of Kounellis using our knowledge and the light sources available.

Balestreri Studio has remained a strong memory of a life's work due to significant interaction with the various professional and artistic involved like Kounellis and Arnaldo Pomodoro with all their staff, Ermanno Olmi who filmed the work in progress every day, the photographer Carlo Orsi etc .

This experience is about how the choice of light for an artwork often can not be interpreted but must follow the choices of the artist or it must allow the artwork itself.

### Lighting and Painting

The most proper lighting of a painting must, instead, transfer coherently what the author wants to tell. The

observer has to be captured by the details that, with the light, can be emphasized and more captured. A completely different choice is made for the exhibition "Gastone Novelli. Retrospective" in which the paintings do not require an interpretive lighting that would allow light to return to the depth brush strokes, therefore has decided to get a light that would allow viewers to catch all the details of the work. The result is a soft, diffused, enveloping light properly carry out its task.

**Light and Graphic**

Sometimes, however, it is necessary simply some special filters to illuminate the graphic that needs to be read in its full two-dimensionality. This has happened to show "Double Dream of the art", where two-dimensional artworks were exposed and the light helped to make them more readable.

米兰波莫多罗基金会展览馆因其有着整体视野而显得十分独特，有别于传统博物馆的局限环境。因为基金会建在Riva Calzoni汽轮机厂里，所以显得非同一般。

建筑围护的大窗户引入了大量的自然光线，所以光元素成为基金会建筑的一大特色。人造光也是一个不错的选择，可以弱化自然光。

强光形成的光布景突显了艺术品，横梁上的灯具展现了墙上的艺术品，同时也将光线打上墙壁。由此，展览品结合与展现出相同的结构，但不同的时间区分了它们。

日间，建筑和展品以各种方式相互辉映，自然光线的魔力笼罩着空间，艺术品是一体的；夜间，人造光美化了艺术品与建筑，建筑本身也成为了真正的作品。

功能极多的摄影机可安装各种辅助设备，如滤色片成型机，抗眩光屏等。

不同功率和半径的金卤灯、卤素灯等光源，为光线的搭配提供条件，营造出不同色温的光场景，进而具备更为灵活的特性来满足不同艺术品的需求，如艺术装置、雕塑、画、照片等。

根据展览空间的容量及其不同展区的特点，展品呈现不同的维度和特色，因此采用可调光的产品。

## 光和雕塑

"1972年-2008年阿纳尔多·波莫多罗伟大作品展"表达了光和雕塑相结合的理念，雕塑展品丰富的种类令人惊叹，也令照明设计师决定采用有别于当代艺术博物馆趋势的设计，使用漫射灯来营造纯净的空间，进而使艺术品不相混淆。

建筑围护结构被视为一个非常重要的元素，装饰着大束的冷色光，与背景形成强烈的对比，使雕塑呼之欲出，不同光感的营造有助于展厅整体布局的设计。

每件艺术品从强光里浮现，卤素灯发出的射光从全角度将所有雕塑包围。

## 光和装置

"詹尼斯单一行为展"表达了光和装置相结合的理念，艺术家詹尼斯·库内里斯采用了一种不同的手法来体现这一关系，向观者展现了与其艺术装置更相协调的光效果。在整个装置展览期间，艺术家日夜都坚守着基金会展览馆这一空间，目的在于使其作品在这个展厅里大放异彩。由此，库内里斯展现了最为奇幻的光效。照明设计师负责利用自己的照明知识和已有的灯源，来阐释库内里斯的理念。

巴莱斯特雷里工作室在工作中留下了终生难忘的回忆，来自与相关专业人士及艺术家的有效互动，如库内里斯、阿纳尔多·波莫多罗和他们的所有员工，以影像记录每日工作的埃曼诺·奥尔米，摄影师卡罗·奥尔西等。

这是关于艺术品的灯光选择经常无法诠释的经

验，必须遵从艺术家或艺术品本身的选择。

## 光与油画

最适合画的照明必须清晰无误地传递画家想要表达的内容。灯光必须能够引起观者对画的细节的关注，能突显细部，增加细部的吸引力。

"贾斯通·诺威利怀旧展"采用了完全不同的灯效。画作并不需要灯光强调的笔触的深度，因此决定制造允许观者捕获作品所有细节的灯光，最终一种柔和、漫射、环绕的灯光恰当地完成了任务。

# 02

## Commercial & Entertainment
## 商业+娱乐休闲

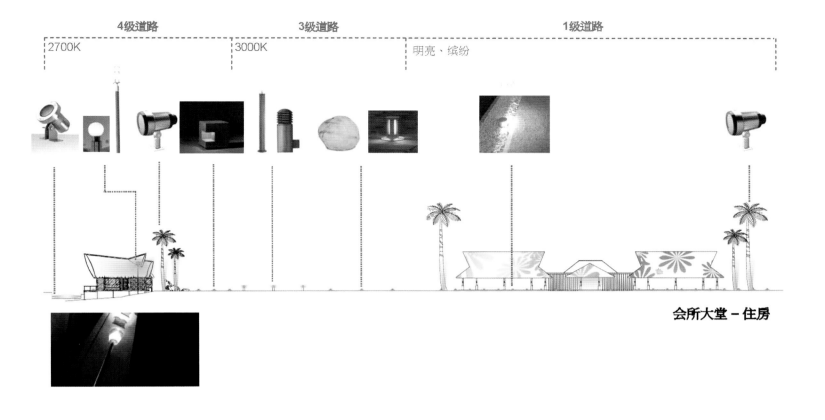

| | |
|---|---|
| ■ ——Saffire Tasmania<br>塔斯马尼亚岛莎菲尔度假村 | The lighting design philosophy was inspired by preserving the environment so that the resort has minimal impact on the landscape after dark.<br>照明设计的哲学基于保护环境，使度假村在夜晚对景观的影响降到最低。 |
| ■ ——The Island Club<br>三亚东锣岛 | The fundamental concept behind the design is to keep the island's original nature and low carbon dioxide emission environment.<br>对于其照明设计赋予的定义为"自然环境的原生、人文休闲的感知"。 |
| ■ ——Surfers Paradise Hilton<br>冲浪者天堂希尔顿酒店 | The lighting design was extensive…from basic forms o f apartment lighting through to relatively complex colour change systems used for the external facade light effect.<br>从公寓内最基本的照明系统到建筑外立面上稍微复杂的变色照明系统，照明设计就是这样无处不在。 |
| ■ ——RED Prime Steak<br>别克大厦红色牛排餐厅 | The rays frame a dramatic procession for diners entering the Main Dining Room.<br>灯光排成戏剧性的队列带领用餐者进入主餐厅。 |
| ■ ——Duke of York Square and Headquarters Building<br>伦敦约克公爵广场及其总部大楼 | Throughout the design, consideration for local residents and the environment was always significant. The challenge was to create an architecturally informed and powerful lighting response whilst maintaining restraint.<br>在设计过程中，考虑当地居民和周边环境总是非常必要的。而设计面临的挑战在于：需要在静穆的状态下体现建筑的内涵和光照的强度。 |

# Saffire Tasmania

## 照明与生态的融合
——塔斯马尼亚岛莎菲尔度假村

**Credits**
Location: Tasmania, Australia
Lighting Design: PointOfView
Architect: Robert Morris Nunn/Circa Architecture
Interior Designer: Chada
Photography: Peter White 7 David Becker
Awards: 2010 IES Vic "Excellence";
2011 IES Australia & New Zealand 'Excellence'

Saffire Tasmania is built on the site of an old caravan Park on the fringe of National Park on the beautiful and remote Freycinet coast, close to the famous 'Wine Glass Bay'. It was the architect's intention that the purity of the environment be reflected in a fresh, modern and pure design style. The interior design was inspired by the natural environment, a pristine coastal landscape; "coastal freedom".

The lighting design philosophy was inspired by preserving the environment so that the resort has minimal impact on the landscape after dark. PointOfView reflected on the quantity of light that had once emanated from the site when it operated as a camping ground, and focussed their efforts on trying to not to exceed the night time impact of a few camp fires. All lighting decisions considered from an exterior perspective; what would be seen when viewed from a distance. Preservation of darkness was paramount in formulating the lighting design to limit detriment to the natural environment. As a result light levels are kept very low and strategically placed; sources are shielded and mostly focussed to the task, and landscape illumination is bare minimum to protect natural ambient qualities and avoid distracting spill light.

**Main Building**
**Arrival; Level 3**

Upon arriving at the Saffire resort, one is greeted by a 50m walkway statement. A key architectural feature is the fluid roof/ceiling form that draws the visitor towards the main building reception. This ceiling is devoid of any penetration for lighting.

The arrival walkway is flanked by columns guiding the visitor towards the reception. In the evening, attention is immediately drawn to this entrance by the illumination of the columns through dichroic filters. Warm white LED is also concealed in details along the entrance boardwalk providing washes of light, and highlighting the colour and materials used in the space. At the end of the entry walk is the reception and waiting lounge where large decorative floor lamps are assigned to each individual table and chair arrangement. Light levels are modest to create an exclusive and intimate impression. Much of the general lighting is integrated into joinery and building details. Warm white LED strips are concealed within joinery units and the ceiling is gently washed from a wall cove detail using high output cold cathode.

A feature stair case that leads down to lower levels lit via small fixtures integrated into niches of the joinery wall on one side

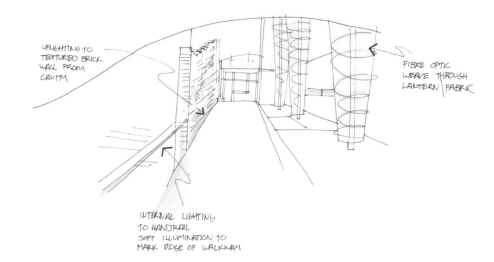

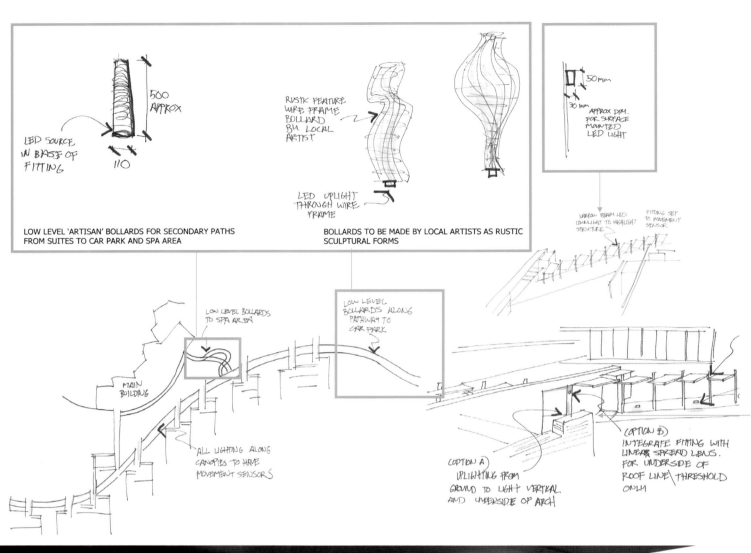

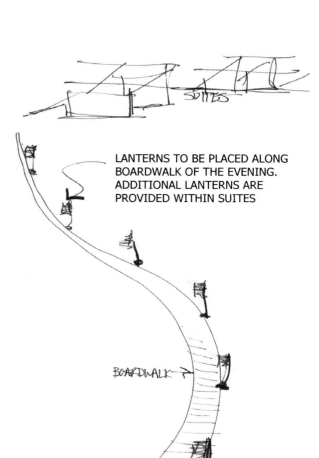

LANTERNS TO BE PLACED ALONG BOARDWALK OF THE EVENING. ADDITIONAL LANTERNS ARE PROVIDED WITHIN SUITES

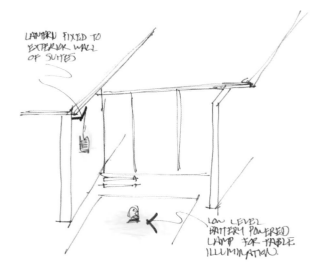

BATTERY POWERED TABLE LAMPS (TAKEN OUT OF LANTERN HOUSING) OR MOUNTED AS A LANTERN IN EXTERIOR AREAS OF THE SUITES.

THE SAME LANTERN CAN BE USED ALONG THE BOARDWALK

LANTERNS ALONG BOARDWALK

providing decorative pockets of light, and task lighting to the stairs.

The other side of the stairs opens to a waterfall. This is lit through the lower tank by submerged halogen up lights through the splash of the falling water.

Level 2

Restaurant

Once in the dining room, customised pendants provide a narrow band of light along the glazing line, and also a gentle wash upwards. The main thrust of the lighting design in this area was to limit reflections in the window that might spoil the fabulous views, and also to limit visibility of the building from Coals Bay. The lights are located at the glazing mullions to assist in this ambition.

Tables in the raised floor area are lit by individual customised table lamps that provide a small pool of light on each table.

Suspended from the double height ceiling above this space are 3 large custom feature pendants that are a dense bunching of acrylic rods, lit by LED. The rods are an artistic topographical representation of the mountains beyond. The rods are housed in a black shade so that the light is concealed, to reduce spill and reflections in the glazing.

Another element of the dining lighting is (as a backdrop) a substantial timber wine wall. This glows to provide ambient light.

ENTRY POINTS FOR DELIVERIES ON SENSOR SWITCHES

LOW LEVEL LIGHTING AT DROP OFF

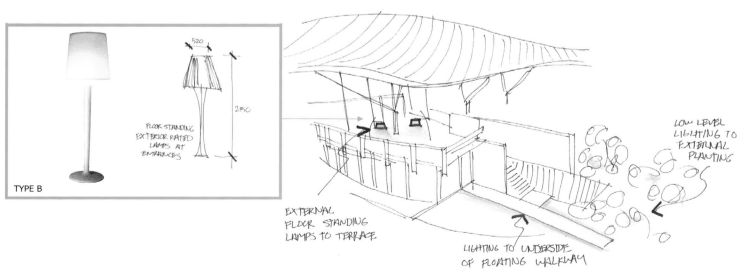

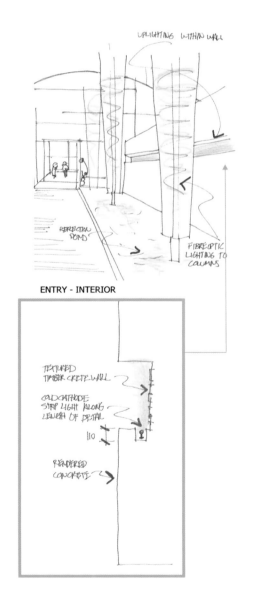
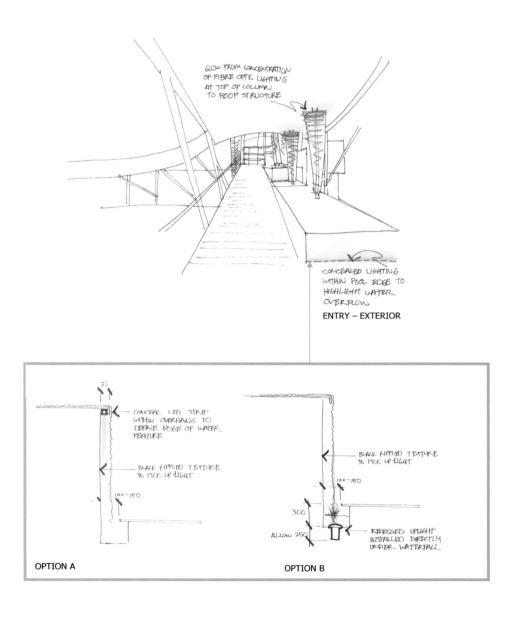

ENTRY – INTERIOR

ENTRY – EXTERIOR

OPTION A

OPTION B

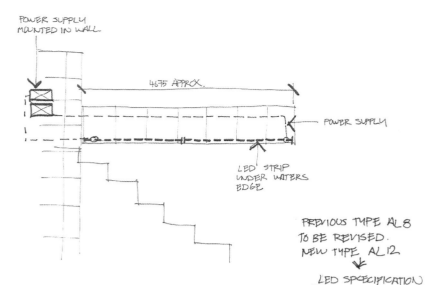
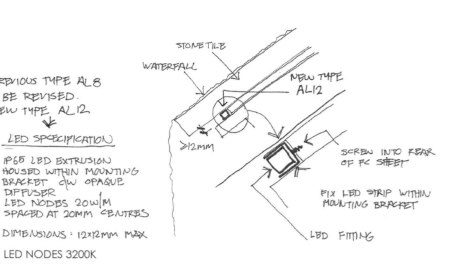

### Lounge & Bar

The adjacent lounge is lit exclusively by individual floor standards that have a focussed projection of light to the seating groups. The bar itself is lit from above by simple black surface mounted adjustable halogen fixtures, supplemented by linear LED fixtures recessed into the bar details.

### Level 1

### Board room & Pre-function

The pre-function space also operates as a gallery for local artists, for which Tasmania is well known. It is lit by semi recessed down lights where the lamp is heavily masked to comply with the rationale noted elsewhere.

The adjacent boardroom is a multi-purpose space, and accordingly combines up lighting from pendants, down lights, decorative wall washing (halogen) and functional wall washing (fluorescent). A control panel system provides a range of scenes to meet the various uses of the room. Lighting scenes are also coordinated via the AMX system.

### Spa

The spa is generally low lit to create intimate mood. At reception, lighting is a mixture of feature floor light and integrated display lighting.

Circulation is lit by a variation of the lift wall light – custom designed wall bracket light. In the treatment rooms lighting is indirect, and separately controllable for various 'scenes'. Flush recessed CFL wall lights provide a soft feature to the walls, with additional illumination from linear fluorescent lighting integrated into the joinery.

### Guest Suites

Guest accommodation comprises just 20 individual, modern and elegantly styled bungalows that fan out below the main building, all with magnificent views of the 'Hazard' mountain range beyond. Great emphasis was placed on maintaining discreet methods of illumination to support the refined architectural treatment and core lighting design principle of minimising impact to the natural environment.

The ceilings in the accommodation are clean of any lighting; general illumination is provided by LED linear source mounted to the top of a main wall on one side, washing the ceiling. Other illumination comes from joinery integrated lights and FFE items. An LED marker light provides night light access to the bathroom. In the shower a

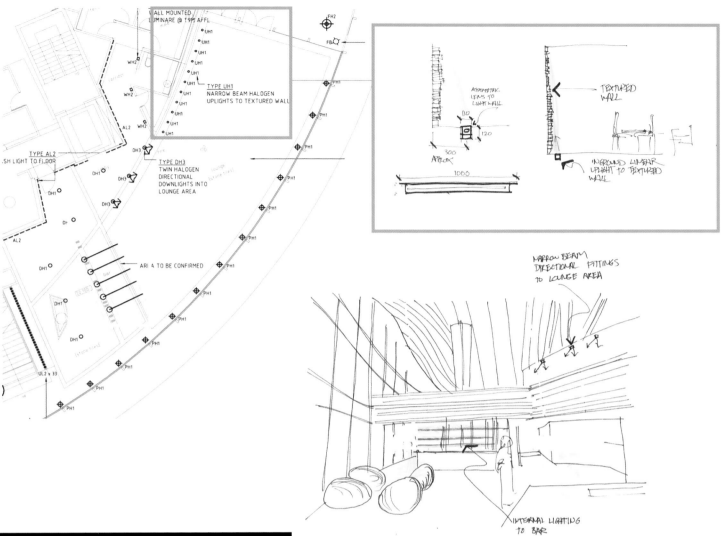

feature is provided by light integrated with the shower head, so light is transmitted through the water cascade. Fluorescent mirror light works with halogen down lights.

Each Suite has a deck. This features a custom designed 'lantern' that integrates an in-ground up light with a decorative wire element. Diffuse light is reflected across the deck, avoiding visible light from the Coals Bay township.

### Exterior & Landscape

Lighting to the exterior is deliberately very limited in order to preserve darkness of the environment as much as possible. With the exception of modest lighting to a small planter bed at arrival (all downcast) there is no feature lighting of planting.

The Suite walkways are illuminated with a single 3W LED at 3m centres recessed into the canopy legs, aimed down. These lights are managed via presence detection. Some additional illumination is provided at each blade (service node) – up light is contained by canopy overhead so there is no waste into the sky.

Custom 'jewel' glass bollards were developed for key access ways into the building. These are glass (9mm) cylinders with crushed glass inside. The fittings utilise a standard in-ground halogen up light (35W IRC) and so the technical aspect is simple and proven.

Complementary walkways are lit by CFL in an asymmetric surface down light.

At the car park, an induction coil in the road activates the car park bollards from vehicular traffic. The bollards provide asymmetric downlight with no upward spill. An illuminated button provides pedestrian activation of these lights.

### Control

The entire lighting system is managed by Philips Dynalite dimmers and control. In the board room the Philips Dynalite syetem interfaces with AMX to provide integrated control of lighting and audiovisual devices in this facility. Scenes are generally time clock or presence managed, with over-ride panels at each area. A master touch screen panel is mounted at reception.

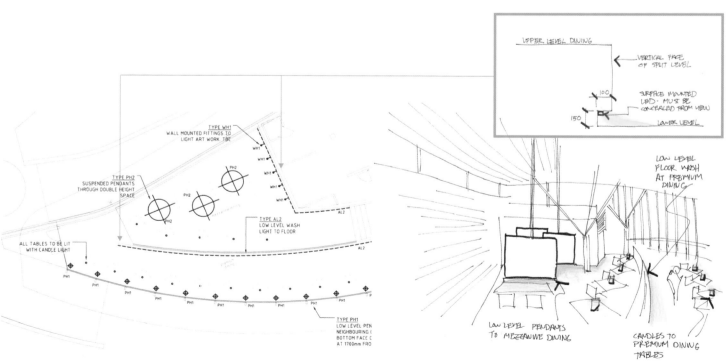

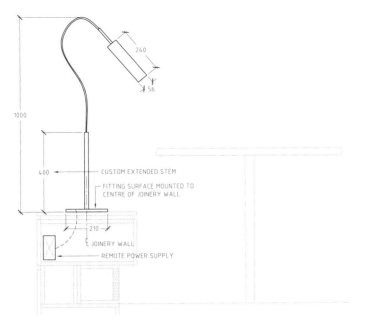
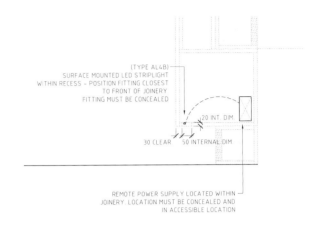

② TYPE TH5 - MEZZANINE TABLE LIGHT    ③ TYPE AL4B - JOINERY WALL STRIPLIGHT

莎菲尔度假村建在国家公园边缘的原旅行车停车场遗址上，毗邻美丽偏僻的菲瑟涅海岸和著名的"酒杯湾"。设计师意在通过新鲜、现代和纯粹的设计风格来反映环境的纯净。室内设计是受原始海岸景观的自然环境的启发。

照明设计的哲学基于保护环境，使度假村在夜晚对景观的影响降到最低。PointOfView公司思考着作为一个露营场地时灯光发射的质量，并致力于使之不超越营火对夜间的影响。所有的照明设计都以外部透视为基准，参考远距离时的影像。保持相对昏暗的环境在于使照明设计对自然环境的影响降到最低。因此，灯光的等级非常低，并且被战略性地安置。发光源隐蔽地藏匿于景观之中，以减少对自然环境的不利影响和避免恼人的点光源。

**主建筑**

入口：3楼

在抵达莎菲尔度假之前有一条50米长的走廊，一个关键的建筑特点是流线型的天花板，指引游客进入主建筑。天花板上没有安装任何照明设备。

走廊侧翼的立柱引导游客进入前台。傍晚，被双色滤光灯照亮的立柱立刻成为了焦点。温暖的白色LED灯也隐藏在入口木板路两侧，突出了这个空间的颜色和材料。

入口末端是前台和休息室，那里的每个桌椅旁都有大型的装饰落地灯。适度的亮度营造了私密的感觉，很多这样的照明设备都结合了细木工艺和建筑细节。温暖的白色LED灯条隐藏在细木工单元里，天花板被镶嵌在墙壁里的高能冷阴极灯照亮。

通向下方的颇具特色的楼梯旁有一面带壁龛的细

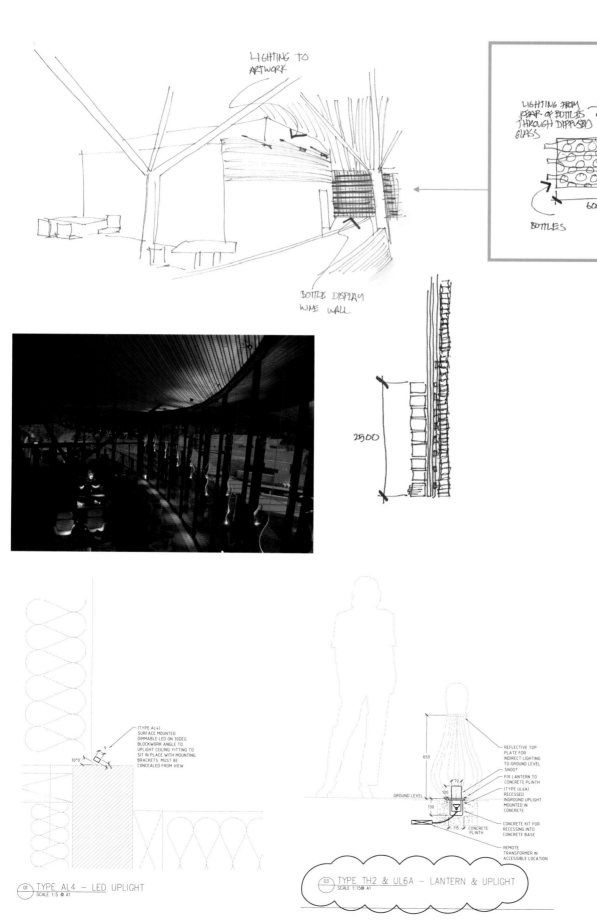

木工墙，里面的小型灯具用作装饰照明和楼梯的功能性照明。

楼梯的另一面面向瀑布，被淹没在下层水槽里飞溅的瀑布水下的卤灯照亮。

二楼餐厅

餐厅玻璃上定制的垂花饰提供了狭窄的光线，优雅地向上弥散。主要的照明设计理念是限制可能破坏这里美好景观的窗户反射光，同时也降低从炭湾方向对建筑的可见度，因此灯具被安放在了窗户的直楞之上。

地面上被抬升的桌子由独立的定制桌灯照亮，在每张桌子上形成一个小的灯池。

3个大型的由许多丙烯酸棒组成的定制垂灯悬浮于双层高的天花板之下，使用LED照明。这些棒体通过艺术表现形式展现了起伏的山脉，并且被黑色灯罩围住，以确保光源隐蔽和减少反射光。

餐厅的另一个光源(作为背景)是一面木质酒墙，提供了环绕照明。

休闲吧

旁边的休闲吧有独立的灯具将光线聚焦在座椅区，这个吧本身由上方黑色外壳的可调节卤灯照亮，并补充有线型的LED灯。

一楼会议室和宴会厅

宴会厅也是当地艺术家长廊，是塔斯马尼亚著名的特色。那里被半嵌入式的灯具照亮，和其他地方一样，也阻挡了大部分的光线。

会议室是一个多功能空间，有下射灯、装饰性的壁灯(卤灯)和功能性

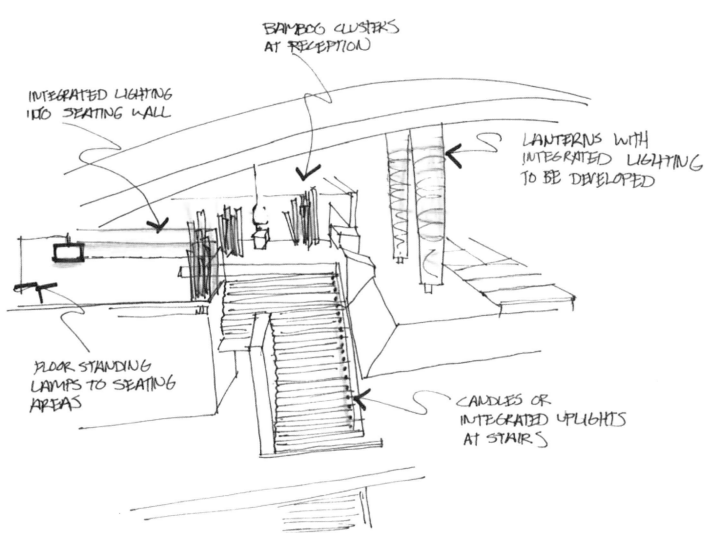

INTEGRATED LIGHTING INTO SEATING WALL

BAMBOO CLUSTERS AT RECEPTION

LANTERNS WITH INTEGRATED LIGHTING TO BE DEVELOPED

FLOOR STANDING LAMPS TO SEATING AREAS

CANDLES OR INTEGRATED UPLIGHTS AT STAIRS

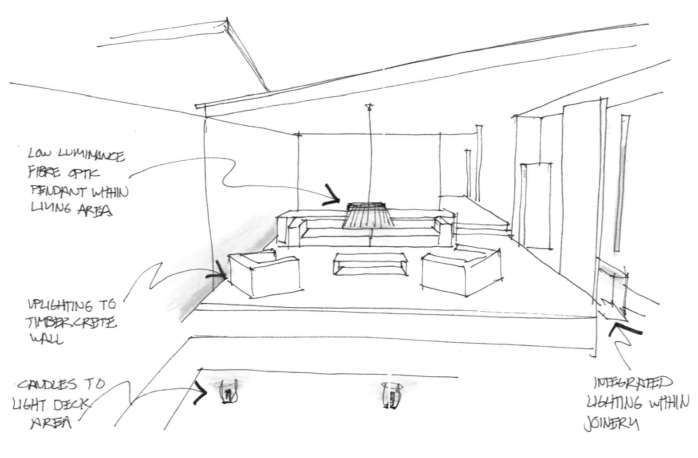
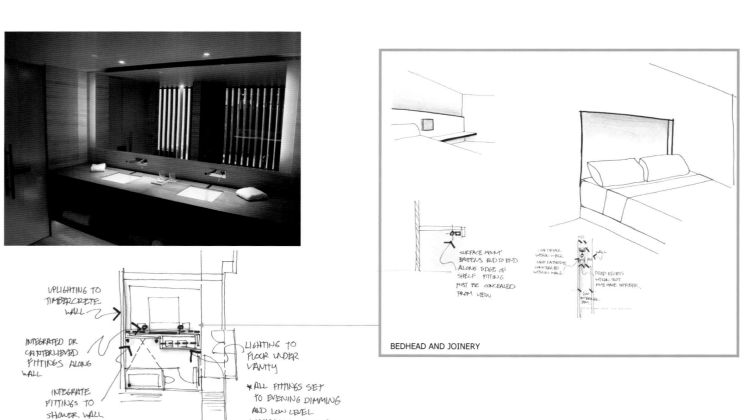

BEDHEAD AND JOINERY

的壁灯(荧光灯)以配合多种用途，由一个控制屏系统控制，还有辅助的AMX系统调节灯光效果。

**水疗间**

水疗间采用低照度灯源来营造私密氛围。入口处有一系列特色楼板灯和整体照明。

楼梯旁有一系列壁灯——定制的墙壁包含了照明设备。房间里的灯光是非直接的，可根据不同的场景控制。明亮的嵌壁式CFL壁灯为墙壁提供了柔和的光线，以及在细木工墙里补充的线型荧光灯。

**客房**

客房呈扇形排布在主建筑下方，由20个独立的时尚优雅的小屋组成，每个都能欣赏到对面哈扎山壮观的景色。

严谨的照明方式始终贯穿这个精妙的建筑，支持把对自然环境的影响降到最低的核心照明设计原则。

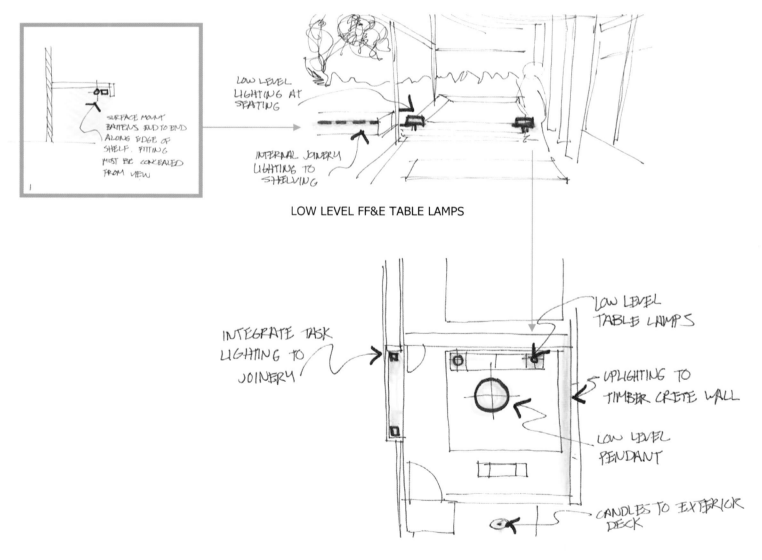

LOW LEVEL FF&E TABLE LAMPS

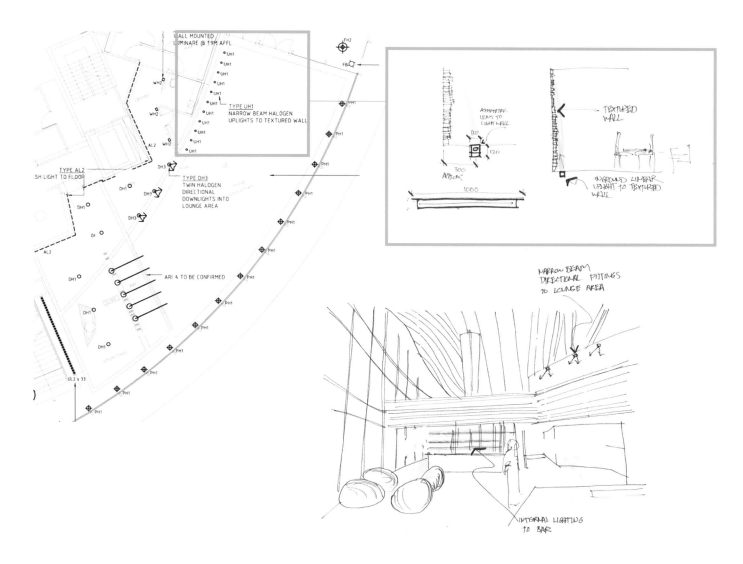

房间里的天花板没有任何照明设备,仅有一面主要的墙壁顶部的LED线型灯提供柔和的光线,并辐射向天花板。其他照明由细木工家装里的照明设备提供。一个LED指示灯给浴室入口提供了夜间指示。淋浴间的一大特色光源与淋浴器头部结合,光线在水中传播。镜面荧光灯和卤素下射灯一起照亮了洗手间。

每个客房都有一个甲板。用装饰性金属丝做成的颇具特色的定制"灯笼"里有一个埋地灯,光线从甲板上弥散开来,避免了从炭湾镇上看到可见光。

外部和景观

外部故意营造的非常微弱的光线为了尽可能地保持环境的幽暗。除了入口下方的一个小花园有微弱的光线之外,植物里没有安装其他照明设备。客房的走道由一个个嵌入天篷脚的3米高的3瓦LED灯向下照射,这些照明设备都是感应的。每个节点有额外的光源,向上的光源被天篷有效地遮蔽了,所以不会射入天空影响环境。

定制的"珠宝"玻璃系船柱式的灯柱是通向建筑的关键,9毫米厚的玻璃圆筒内部是碎玻璃。这

个设备采用了标准35瓦埋地卤素上照灯,这种技术既简单又实用。

其他走道由荧光灯从不对称的表面照射。

停车场里有一个路面感应器来接收来自汽车的信号。接到信号后会启动不对称的下射灯,并且不会向上弥散。一个照明按钮能使路人来控制这些灯。

**系统控制**

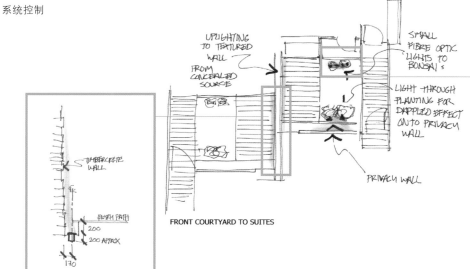

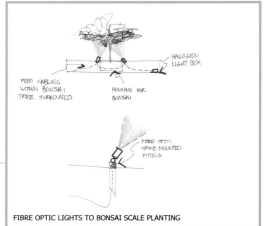

整个照明由Philips Dynalite系统调节控制。会议室里的Philips Dynalite系统辅以AMX系统为这个设施提供了照明和视听设备的总体控制。通常由时钟或感应控制,每个区域有控制面板。前台有一个主触控屏。

# The Island Club

## 岛屿的夜间表情
—— 三亚东锣岛

**Credits**
Location: Sanya, Hainan, China
Scale: 116,000 m²
Client: The Island Club – Dong Luo Dao, Sanya, Hainan
Lighting Design: GUANG Architecture Lighting Design
Architect: Wilson Associates, Shine Design
Photography: GUANG Architecture Lighting Design
Light Source: CFL, CDM, LED
Cost: RMB11,000,000 (USD$1,734,600) (Lighting only)

'The Island Club' is a project to develop a desert island which is situated in South China Sea nearby Sanya, Hainan, Mainland China to a luxury holiday villa. Sanya, Hainan recently has become one of popular holiday paradises in Asia and the name of desert island is called 'Dong Luo Dao'. Dong Luo Dao is an international standard ecological island which is full of natural resources, soft sandy beach, and surrounded by the sea. It has a typical tropical climate and the ocean nearby is as marine ecosystem which is rich in a variety of corals and tropical fishes. Dong Luo Dao is located at 60km away from the centre of Sanya City and the measure of island is approximately 116,000m². The idea of design is to create a mysterious, unique, and having strong image for the island. The fundamental concept behind the design is to keep the island's original nature and low carbon dioxide emission environment. The lights were combined to Dong Luo Dao's distinctive geological structure and diverse natural resources, and bring in Eastern fashion style as the club theme through the light.

**The Grade of Path**
The light scheme for landscape is planning different luminaires, colour temperature, and brightness based on the classifications of the road. The pathways on

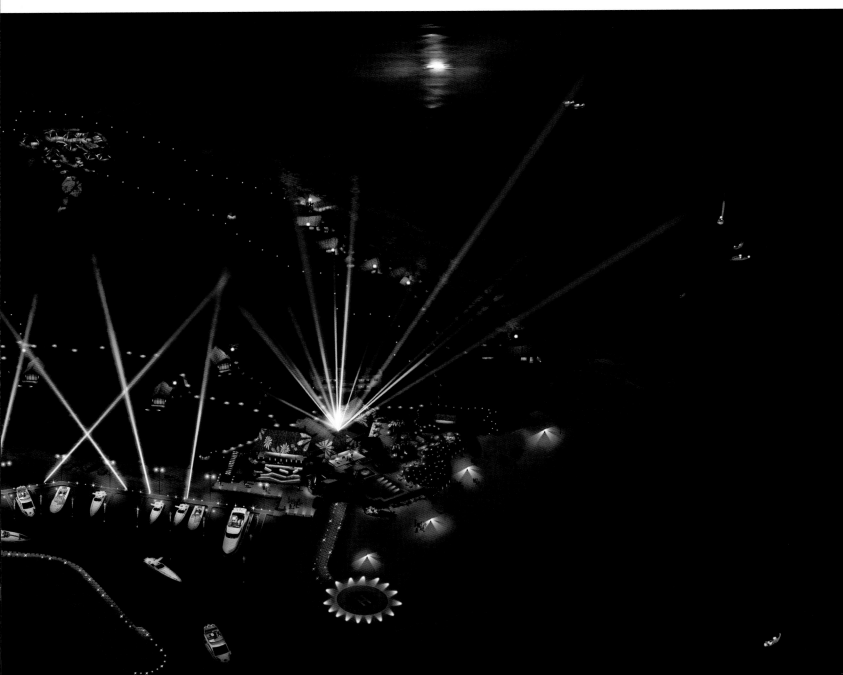

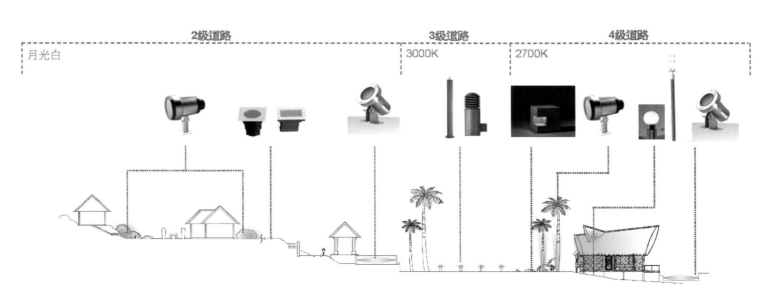

| 2级道路 | 3级道路 | 4级道路 |
|---|---|---|
| 月光白 | 3000K | 2700K |

**Spa区 – 住房**

## 通道照明层级

1. 必经动线
2. 主题引导动线
4. 进入私领域动线
3. 休闲散步动线

一级照明通道
二级照明通道
三级照明通道
四级照明通道

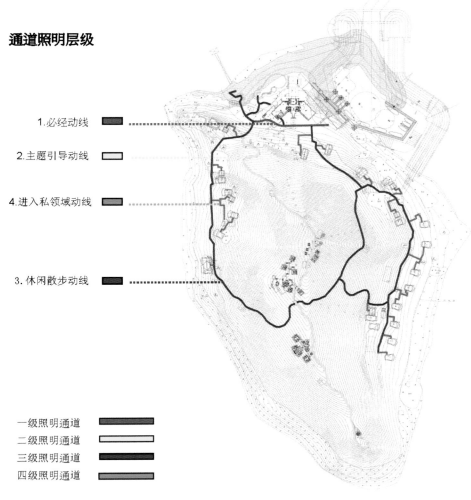

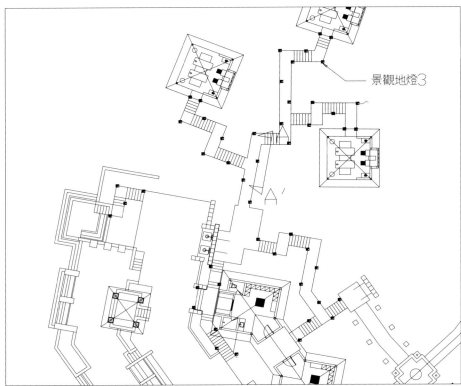

景觀地燈3

水療區木棧道燈具平面配置圖

the entire island can be classified into 4 grades. Each grade of pathway has individual theme and perform lighting with different feelings. First grade is primary passageway that most people consist of clients and island staffs to go through. It is a road to connect lobby building and other public areas. The lighting concept is to create a bright, colourful, and alive impression to the environment. LED Spotlights are arranged beneath palm trees and illuminate down to up. Footlighting were concealed at two sides of pathway with rhythm to indicate direction by floodlighting on wood deck.

Next class of road is a transitional pathway which connects public areas to individual villas. The lighting scheme is to indicate the direction of villa for the residents by planning 4,000K colour temperature recessed floor lights. And, after the transitional area, the path is connecting to fourth grade and having different light source. The idea behind the design is to make a sense of relaxing and calm through changing the quantity of light and its colour temperature which is having 2,700K colour temperature and reduce the brightness of luminaires for creating a silent private zone to the villa residents.

Unlike other three grades which are as a series of lighting plan, third grade of pathway is for leisure walking. 3,000K floor luminaires were planned and the brightness was reduced to lower level. This is in order to let people enjoy the natural night view of Dong Luo Dao, but still keeps safety at the same time. In order to response to the idea of Eco Experience, LED and CDM luminaires were used as major light sources, and monitored by central intelligent lighting control system. Lighting control system benefits the efficiency of luminaires and enhances the lighting experiences.

Lightings on Dong Luo Dao is emphasizing the nature of island through simplified the type of light sources and luminaires style. It focused on planning lights with different methods and integrating light with leading technologies to create comfortable and relaxing ambiances to the island. And, people on the island can interact with lights and enjoy a brand-new lighting experience.

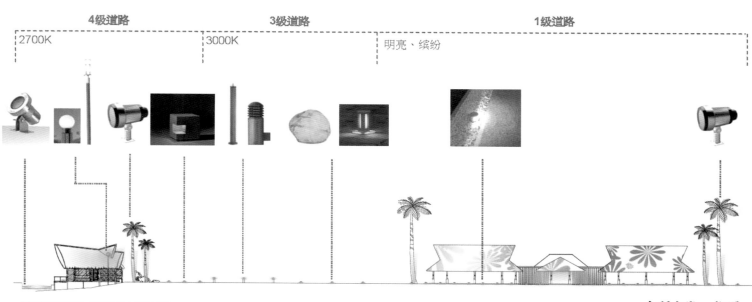

会所大堂 – 住房

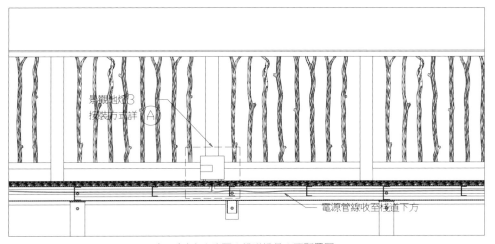

A-A'向水療區木棧道燈具立面配置圖

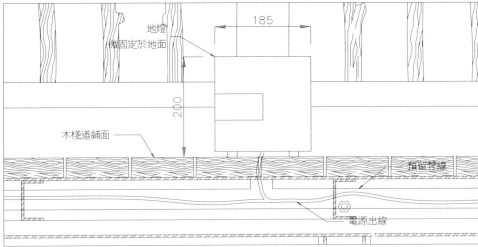

Ⓐ 地燈燈具按裝細部圖

本案为中国海南三亚市西南海上的一座国际标准化原生态岛屿，距三亚市中心约60公里，陆地面积约11.6万平方米，拥有完整且独立的陆、海、空立体空间。岛上自然景观资源非常丰富，为典型的热带雨林气候，海水清澈、沙滩柔和，环岛海域生长着大量的珊瑚，聚集着各式各样的热带鱼类，宛如一巨大的海洋生态圈。

希望带给世人一种神秘、个性、独一无二形象的东锣岛，以打造一原生、低碳环保的顶尖岛屿度假酒店为其市场定位；对于其照明设计赋予的定义为"自然环境的原生、人文休闲的感知"，架构于岛上的原始地貌及丰富的自然资源，以时尚南岛的东方情怀与原生大地的真实况味作主题的导入，利用"光"来映衬其瑰丽动人的一面。

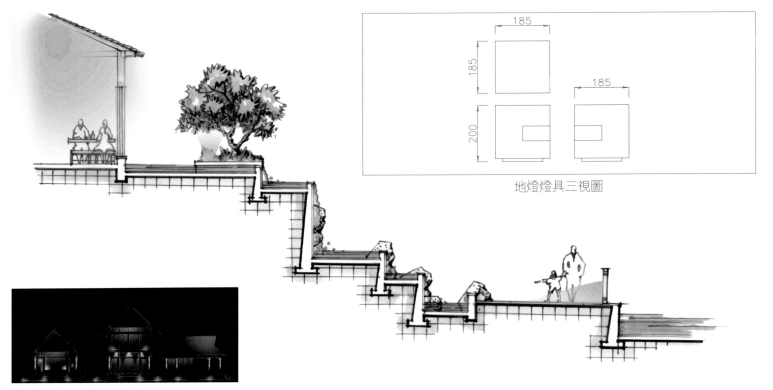

地燈燈具三視圖

137

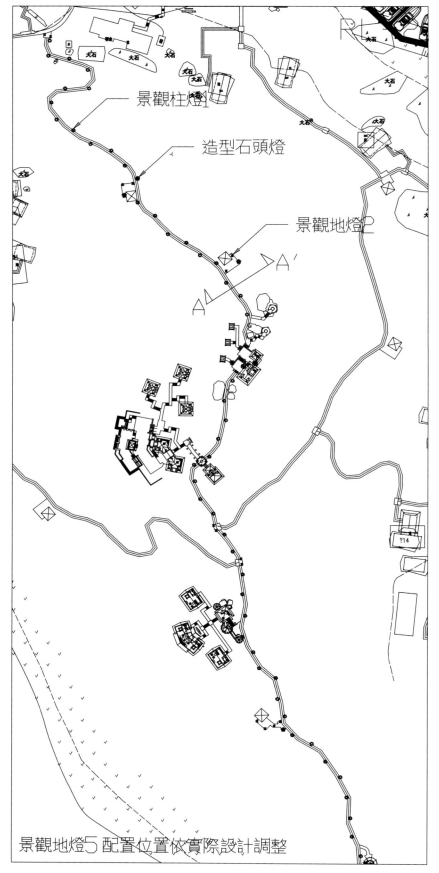

主題引導動線燈具配置圖

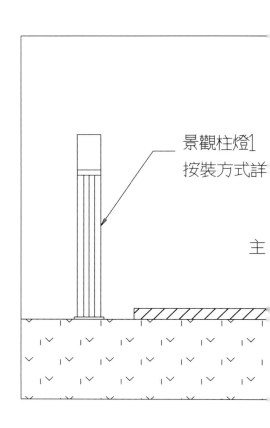

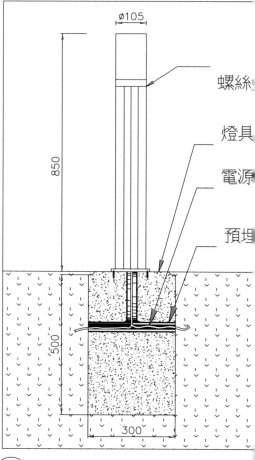

(A) 景觀矮柱燈1 按裝細部圖

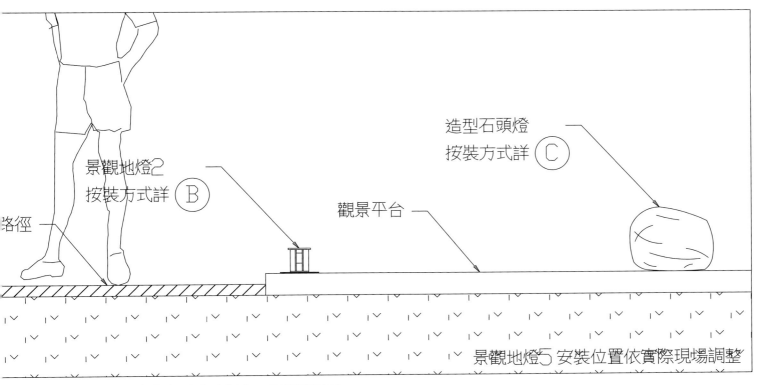

A-A'向路徑燈具立面配置圖

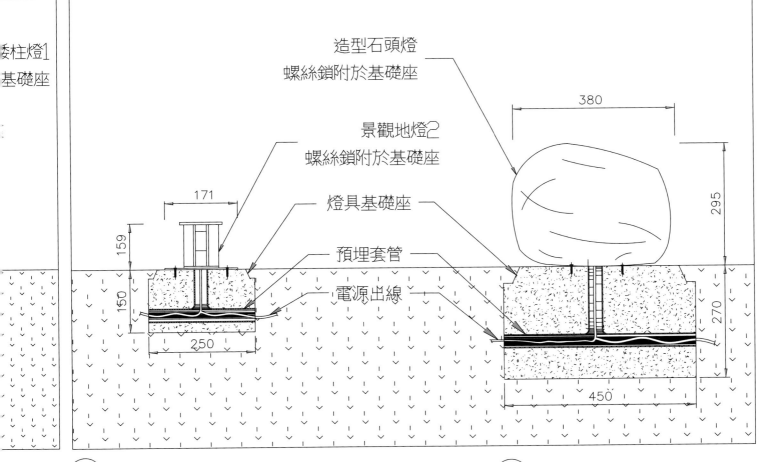

B 景觀地燈2 按裝細部圖

C 造型石頭燈按裝細部圖

# 景观照明规划 | Landscape Lighting

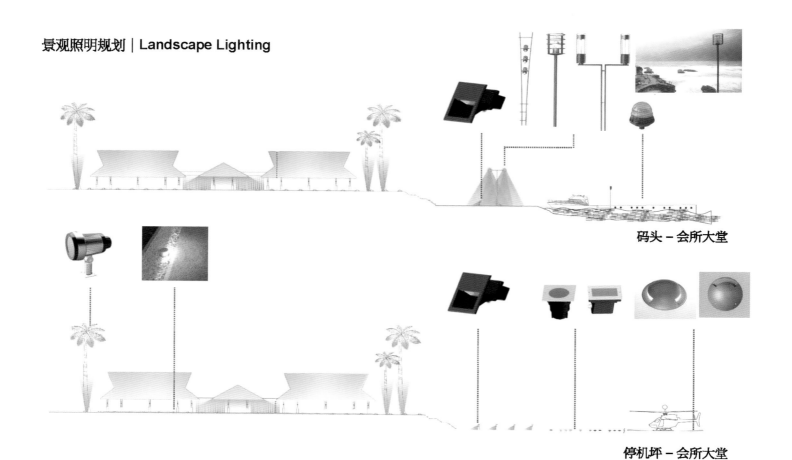

码头 – 会所大堂

停机坪 – 会所大堂

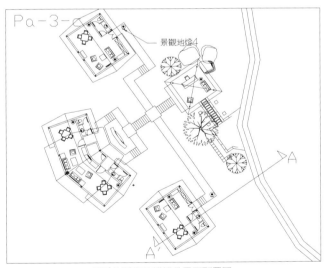

棋牌室景觀地燈燈具平面配置圖

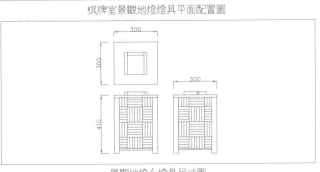

景觀地燈4 燈具尺寸圖

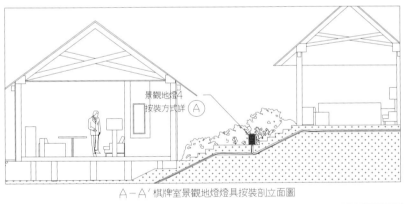

A-A′棋牌室景觀地燈燈具按裝剖立面圖

A 景觀地燈燈具按裝細部圖

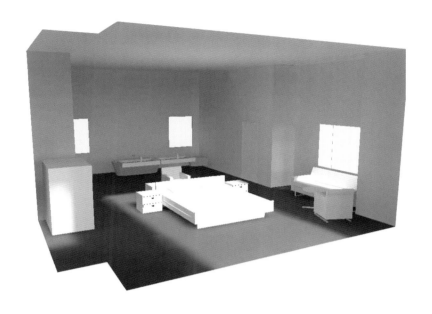

8：00

等照度色阶示意图　　　　　等辉度色阶示意图

(lx)　　　　　　　　　　　(cd/m²)

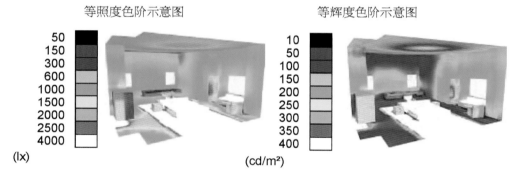

8：00

等照度色阶示意图　　　　　等辉度色阶示意图

(lx)　　　　　　　　　　　(cd/m²)

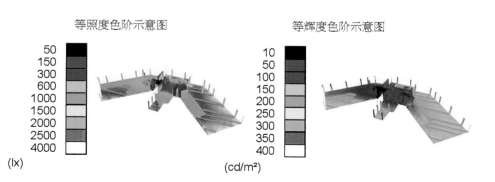

景观照明依不同的尺度作为其布光重点，夜间整体环境沉静、低调，四面环海的开阔夜空中繁星璀璨、与月争辉。凭借其天然的景致，整岛以保有适当的指引性且不造成天空漫射辉光等光污染侵扰为主；通道采取道路照明分级制，可细分为4个等级，依其使用层级的不同来规划个别照度及色温的变化：一级道路连接会所与公共区域，为使用者必经动线，主要创造一种明亮、缤纷的意象；配合具有展演性质的有色光作适度点缀，丰富岛屿夜间的表情。二、三级为公共区域通往住房的过渡道路，采用3 000开暖白与4 000开的月光白色做引导，呈现其柔和、舒适之质感；过道途中提供休憩的节点与木平台，布置富有光影变化的地灯，增添其趣味性。四级道路为延伸至住房的小栈道，使用2 700开色温并降低其亮度，营造一种温暖、静谧的氛围，减少光的侵扰，让住客能够充分享受到东锣岛的夜间样貌。

岛上灯光使用人性化且精准的智慧型灯光控制系统，根据时段变化与人的活动来调整情境，兼顾使用者的视觉享受与安全性；景观灯具使用LED、CDM复金属等节能光源，落实对于Eco Experience原生态的体验。其外形则以植栽、原生岩为其元素，形式简洁，保有中性的基调，朝家族化系统发展并与生态环境融合，延续原生大地的自然纯粹，表达平静沉稳的质感创造。

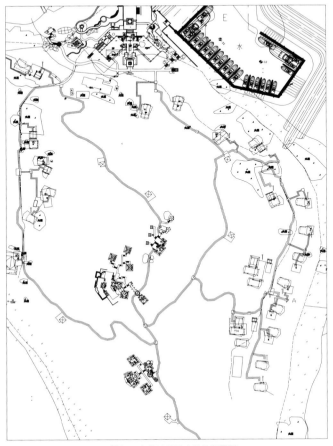

RL-4 茅草頂暈光燈具配置圖

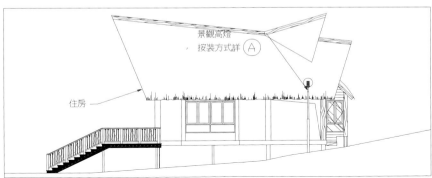

A-A'向景觀高燈立面配置圖

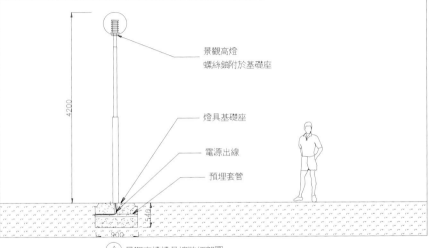

A 景觀高燈燈具按裝細部圖

# Surfers Paradise Foreshore

## 线性半导体特色照明
——冲浪者天堂海滩改造

**Credits**

Location: Australia

Lighting Design: Tony Dowthwaite Lighting Design

Surfers Paradise, located on the Gold Coast, Queensland, Australia is widely regarded as one of the country's premier international tourist destinations.

The attraction being predominately here are the weather and the beachfront.

The central Surfers Paradise area comprises high density, multi storey apartment buildings that border the beachfront.

Recently the Gold Coast City Council embarked on an upgrade/redevelopment of the Surfers Paradise Foreshore.

The entire approx 2 kilometers of foreshore was completely demolished & remodeled to create a wide open space for recreation, pedestrian & "personal movement" use. This essentially comprised wide, spacious walkways/bikeways combined with softscape & amenities. The urban architectural design was by PLACE DESIGN, a well renowned design company in Australia with offices in China.

An important aspect of this renewal was the lighting systems. Precincts such as this are used extensively at night. They also host regular night markets & are used for yearly events such as New Year's Eve celebrations. Large crowds & public safety were key issues when developing the light planning for the site. Our approach was to categorize the light design into three areas...

**Amenity Lighting**

Amenity Lighting is basically the area lighting that achieves the required illumination levels in accordance with the relevant Australian Standards. Here we chose 11m high mast fixtures to support the area lighting for both the pedestrian zone & the adjacent carpark & roadway (The Esplanade). The luminaries house 400W metal halide sources with forward throw distribution reflector systems for pedestrian side & elliptical roadway distribution for the road. These fixtures are located on what is termed the *A line. Housed in the base of these lighting

masts are the low voltage supply/driver devices for the various LED systems in the "sacrificial" zone, both Architectural & Feature Lighting systems...

**Architectural Lighting**

The architectural lighting is more specific...Attention was focused on built elements such as low level seating that is scattered throughout the site & the stainless steel balustrade used extensively through the precinct.

**Feature Lighting**

The feature lighting category utilizes techniques to create visual interest & movement. In this project the linear LED systems, randomly placed throughout the precinct are combined with selective highlighting of planting & 3D objects.

The Gold Coast, in the past, has been subject to cyclones & as a consequence, built, man made things, along the foreshore are required to be

"sacrificial", that is, everything east (on the ocean side) of a predetermined line...the A line...has the potential, during extreme weather events, to be destroyed. Any mains power systems, 240V or greater are not permitted in the area.

In addition to the area lighting beach volleyball courts are specifically lit with high performance floodlights for night competition play.

The effect of the lighting design results in an extensive public precinct that combines area lighting, color & "sparkle" to create visually interesting outcome.

冲浪者天堂位于澳大利亚昆士兰州的黄金海岸，这里是澳大利亚最顶级的国际旅游度假胜地之一。
气候和海滩是这里的两大特色。
冲浪者天堂的中心区域内主要坐落着高密度的多层公寓楼，与海滩相邻。
现如今黄金海岸市政局着手对冲浪者天堂海滩进行升级改造。
约2公里长的海滩将被彻底改造成一个开放的娱乐空间，供行人和个人活动使用。空间内建有宽敞的步行走道、自行车道，并配备有软景观及便利设施。城市建筑设计由普利斯设计集团设计事务所负责，他们在澳大利亚家喻户晓，并在中国设有办公室。
照明系统是本次改建的重点。海滨区域在夜间使用频繁。那里也有常规的夜市，并举办诸如迎新年庆典之类的年度盛会。在规划该场所的照明时，重点需考虑大客流群体的安全问题。设计师将灯光设计分为三个部分：

便利照明

便利照明通常指区域照明，亮度根据澳大利亚照明标准确定。设计师选择了11米高的桅灯，使该区域照明设施覆盖行人区和对面的停车库与马路（广场）。照明体为400瓦的金卤灯，而人行道侧装有前照反射器，道路上安有椭圆面反射器。这些装置均位于预设的"A"线上。灯杆底部装有低功率的电源及控制装置，为"牺牲带"多样化的灯照系统提供服务，包括建筑与特色照明系统在内。

建筑照明

建筑照明相对更为明确，主要集中于建造元素上，诸如海滩各处的矮座椅以及照明区广泛使用的不锈钢栏杆等。

特色照明

特色照明通过运用高科技手段，完美彰显视觉和动感效果。在这套系统中，区域内罕有的线性半导体照明系统与一旁精心挑选的植被、三维饰物相得益彰。

过去，黄金海岸时常遭遇飓风的袭击。因而所有海岸线边的建筑物和人造物注定成为"牺牲品"，换言之，在极端气候下，位于预设线——"A"线——东侧的一切都有毁坏的风险。

该区域没有任何电压240伏及以上的大型电力系统。

除了区域照明系统外，为了保证夜间赛事的进行，沙滩排球场安装了高性能的泛光灯。

灯光设计使公共区成为了五光十色、"璀璨耀眼"的"照明区"，从而营造出动人的视觉效果。

# Surfers Paradise Hilton

## 变色照明的视觉体验
——冲浪者天堂希尔顿酒店

**Credits**

Location: Australia

Lighting Design: Tony Dowthwaite Lighting Design

The Hilton Hotel in Surfers Paradise completed November 2011 comprises two towers, a recreation/spa deck above a retail precinct at ground level.
The west tower houses apartments which are a combination of owner occupied & Hilton managed properties.
The taller 55+ level east tower is a mix of hotel rooms, apartments, high end sky homes & penthouses, again all managed by the Hilton group.
Below the east tower are the usual foyer spaces of a 5 star hotel including restaurants, bars, meeting rooms & function spaces.
Our company was engaged to design the lighting systems for the entire complex.
The architectural design was by The Buchan Group, an Australia wide practice with offices in most cities including the Gold Coast. They were also the interior designers for the majority of the FOH areas.
The bar & restaurant on Level 1 were the

responsibility of a Sydney based design practice LANDINI & ASSOCIATES.
The landscape architects were EDAW, again a well known practice with offices in Australia & beyond.
The lighting design was extensive...from basic forms of apartment lighting through to relatively complex colour change systems used for the external facade light effect.
Our role in projects such as this is to understand principles of the lead designers & combine these ideals with the brief & instruction from the client & the needs of the operator, in this case Hilton. It is a complex process that involves many meetings & discussions to arrive at conclusions & ultimately the built form.
As with all new development (commercial & residential) in Australia energy issues are the basis of (our) design approach. The days of inefficient light sources are behind us & although there is a place,

we believe, for halogen sources in selected areas of these properties, the target light sources are either LED or considered use of either compact or linear fluorescent. Where sources such as fluorescent are considered in concealed ceiling coves/coffers where the source is not visible.

The circulation spaces around the lift core of the building & breakout spaces off the function rooms employ this concealed light system at the ceiling/wall junction.

The Lounge Bar & restaurant on Level 1 has a high level of interior design associated with the spaces. Our role was to understand the requirements of the designers & help realize the outcome through assistance with fixture selection, placement & lighting control, an important aspect of any front of house area of a hotel.

Whilst the interior lighting design has a direct link to the interior concepts the external lighting is a

little more animated. Given the Gold Coast is the "playground" of the Australian tourist market we chose to add life to the lower "pedestrian zone" of the property. Linear colour change LED systems are used on the entrance facade to provide a visually exciting experience for the guests & public alike. Continuous linear slots with concealed neon define the arc of the retail precinct at ground & first levels. A feature in the lobby zone is the stair to level 1, it enclosed by a metal cage that floats off the walls & ceiling...Recessed lensed LED down lighters project sharp beams of light through the cage producing a random pattern on the adjacent wall surfaces. Considered uses of LED projectors create a subtle wash of light to shade battening above the retail arcade.

The recreation deck on Level 2 utilizes a number of different lighting techniques including high intensity fixtures to the main lap pool combined with coloured LED in the underwater seating zones along one side of the pool.

Selected trees are highlighted via ground mounted low wattage metal halide sources.

In ground LED strips are selectively placed to provide visual interest throughout the circulation areas of the zone.

Strategically located on this level are a number of high powered filtered (blue) floodlights that subtlety feature the towers at night.

155

位于冲浪者天堂的希尔顿酒店于2011年11月完工。酒店包含两座大楼，以及位于底层零售店铺上方的休憩平台与桑拿浴场。

西侧大楼内为公寓套房，部分被业主占用，部分归希尔顿管理。

东侧较高的大楼超过55层，内建有酒店客房、公寓与高端楼顶别墅，归希尔顿集团管理。

东侧大楼底部是五星级酒店中常见的大堂，内设餐厅、酒吧、会议室及其他功能空间。

托尼·道斯威特照明设计事务所负责为整幢酒店设计照明系统。

酒店的建筑设计由拜肯集团完成。该集团在澳大利亚包括黄金海岸在内的大部分城市设有办公室。前堂的大部分区域的室内设计也是由该集团完成的。

一楼的酒吧和餐厅由来自悉尼的兰蒂尼及合伙人设计公司完成。

易道设计公司负责景观设计，该公司也享誉澳大利亚及世界各地。

从公寓内最基本的照明系统到建筑外立面上稍微复杂的变色照明系统，照明设计就是这样无处不在。

在项目中，照明设计师需要了解主设计师的原则，将其与客户的要求、希尔顿经营方的需求相结合。这是一个相当复杂的过程，需要众多繁杂的会议与讨论来使建筑最终成形。

能源问题是所有澳大利亚新开发项目（商业区和住宅）的焦点，同样也是我们设计的出发点。虽然在建筑的某些区域依旧选择性沿用卤素灯，但我们认为，低效照明灯具的时代已经一去不复返。我们需要的要么是LED灯，要么是紧凑或线形的荧光灯。而荧光灯通常安装在

157

天花板凹槽或镶板处，那里不易引人注目。

在大楼电梯四周的通行区域，以及位于功能空间旁的休息室中，照明系统都隐藏在天花板、墙角中不易察觉的地方。

位于1层的高级酒吧和餐厅配备有高规格的内部设计。照明设计师需要领悟设计者的意图，通过灯具的选择与布局，以及照明控制来实现他们的诉求，这对任何酒店的前堂来说都是至关重要的。

内部照明设计与室内设计风格息息相关，而外部照明则显得更为活泼欢快。因为黄金海岸是澳大利亚旅游市场的"游乐场"，我们选择使大楼低处的"行人区"更具生机与活力。我们在入口立面处使用可变色的线性LED灯，使得顾客和公众享受到激情四溢的视觉体验。

安装着霓虹灯的绵延状墙壁槽勾勒出底层和一楼的零售区的轮廓。

通往1楼的楼道是大堂区的一大特色，楼道四周

围绕着金属笼，笼子悬浮在天花板和墙壁上……LED灯的凹透镜面通过笼子将炫亮的灯光反射到邻近的墙壁上。

设计方考虑利用LED投影机，将光线投射在位于零售店拱廊上方的补隙板上。

在二楼的休憩平台上，不同的照明技术汇聚一处，包括高强度照明灯具，小型健身游泳池配备有彩色LED灯，安装在泳池一侧的水下座位区。

在低功率的接地式金属卤素灯的照射下，树木成为公众的焦点。

地面上的LED灯带为通行区域带来新奇的视觉效果。

而楼层内，还别出心裁地安装着数盏高强度（蓝色）泛光灯，在夜晚，它们成为了一道独特的风景线。

# Show Villa at Sanctuary Falls

## 灯光、景观、建筑三位一体
——圣地瀑布度假别墅

**Credits**
Location: Sanctuary Falls, Jumeirah Golf Estate, Dubai, U.A.E.
Scale: 550 m²
Client: Shaikh Holdings
Lighting Design: Studio Lumen Lighting Design & Consultancy
Interior Design: HBA International
General Contractor: Arabtec

Designed by world-class architects, inteior designers and landscape planners from the hotel and resort industry, the Sanctuary Falls Show Villa is a truly unique creation. Located off the center of the main artery inside Sanctuary Falls, it draws direct inspiration from elements within the community such as water, and earth to provide a true "resort home" experience. The clients aim was to deliver the perfect resort home to each of the residents, and one that reflected utmost attention to detail.

In keeping with the client's vision for the property, the main objective of the exterior lighting for the project was to enhance the resort style feel. There are three design styles for the villas across the development, modern, contemporary & traditional. The show villa, was representative of the modern style villa, and a

very integral part of the objective was to accentuate this feature of the building.

The architects and the landscape designers, responded perfectly to the design brief of creating a "one of its kind" residential destination, and drew inspiration from the elements within the community such as water (various lakes, waterfalls, arrival water fountains, swimming pools) and earth (the golf course which hosts the Dubai World Championship, nature views, tree canopies, forested pathways) to provide a true "resort home" experience.

The exterior lighting design began with establishing the main objectives. The first of these was to define a clear hierarchy among the various elements to be accentuated. It was agreed with the client that the main focal point of the façade was to be the main arrival, and the landscape area around the swimming pool. The textured stone panels on the façade/columns came next and the lighting to these had to be balanced with the interior lighting which would filter out through the openings. It was agreed that all the lighting equipment used would be discreetly concealed or be decorative in nature such that it became an integral feature of the overall theme of the exterior, and that all lighting equipment would be dimmable to allow for greater flexibility in control of the final effect.

In keeping with the vision established for the property, that of a resort home, the lighting design scheme for the villa had to be one with a controlled accent. To emphasize the resort style identity of the project, we developed a lighting scheme with a very high accent light vs ambient light ratio, a technique very widely used in hotels and resorts. This was a very tricky thing to do, as considering that this is a home, comfort and a sense of tranquility were very important. Key elements of the building were carefully selected like the textured stone finish on the columns, pergola structures, water features, planting, etc. to create lighting accents. General lighting was kept to minimum and only confined to areas, such as outdoor terraces & balconies, outdoor dining decks, etc. which required a certain level of illumination for functional reasons.

A careful study was carried out of the villa vicinity to understand the impact views towards the villa, to ensure that all visual elements were being addressed in harmony with each other. It was very important for the exterior lighting to be controlled so that views from within the villa looking out, were not compromised. To ensure this, all the lighting fixtures

were completely recess mounted, and the ones which had to be exposed, were installed away from the sight lines, to achieve visual comfort.

The control of all the lighting goes back to the home automation system, designed and developed for the villa. The complete lighting set-up is controlled by this intelligent system, with scenes having being pre-programmed to suit different needs of the home owner, like—"party", "everyday evening", and an "away" scene. This achieves a very carefully controlled lighting environment and also assists in being sustainable, as most of the lighting is dimmed, to reduce the connected load. An interesting interface between sound and light is achivied within the home automation system, such that when the home owner goes into "party" mode for the lighting, a pre-set music is also turned on thereby creating the perfect mood. This is of course set as per the needs and likes of the home owner, but is demonstrative of the effort the team has taken to enhance the experience of the home owner.

圣地瀑布的这栋度假别墅极其标新立异，由来自酒店和旅游业的世界级建筑设计师、室内设计师和景观规划师联合设计完成。别墅远离圣地瀑布的中心区域，从水和大地等元素中汲取灵感，从而提供真正的"度假胜地"体验。
客户意欲将完美的度假胜地呈现在每一个居住者面前，他们对细节十分讲求。
为了打造出客户梦想中的别墅，外部照明力求烘托别墅的度假风情。在别墅设计的发展过程中，三种主流的设计风格分别是现代、当代和传统。这栋度假别墅是现代式别墅中的代表，设计目标中的一个重要方面便是突出别墅的这一风格。
建筑和景观设计师们完美地迎合了设计宗旨——创造"独一无二"的度假胜地。他们从别墅所在社区的水（不同的湖泊、瀑布、入口喷泉以及游泳池）及大地（举办迪拜世界锦标赛的高尔夫球场、自然风光、树冠、森林走道）等元素中汲取灵感，从而让人获得真正的"度假胜地"体验。

外部照明设计始于设定主要目标。首先是明确需要突出的各种元素的主次。设计方与客户达成一致，确认别墅的立面需以入口处、泳池周围的景观区作为主焦点，外立面和柱子上的纹理结构作为第二焦点。外部区域的灯光需要与渗透出的内部灯光相平衡。照明灯具或隐藏起来，或形成自然装饰，使其成为别墅外部的重要特色，照明设施将是可调光的系统，具有极大的灵活性，从而制造出变幻多彩的灯光效果。
与度假别墅的设计重点相同，别墅的灯光设计也注重营造"度假胜地"的氛围。为了强调此风格，照明设计时特意提高了强光和环境光的比率，这种方案在酒店和度假村使用十分广泛。但这也是需要技巧的，因为这里毕竟是居住区，舒适和宁静也同样很重要。大楼的重要构件是精心挑选出来的，如圆柱上的石刻花纹、藤架结构、水景、植被等，这些元素是照明的重点所在。通用照明范围最小，仅限于室外走廊和阳台、室外用餐平台等需要功能性照明的区域。

设计方认真研究了别墅周边，了解景观对别墅的影响，保证所有视觉元素都互相和谐共存。控制外部照明十分重要，因为这样不会损害到外部景致。为此，所有照明灯具全数嵌藏起来，部分显露在外的灯具也都安装在观景视线之外，创造出舒适的视觉效果。
别墅配备的家庭自动化系统则控制着所有灯具的照明。这套智能系统控制着全套照明组合。为了迎合业主的不同需求，设计方预先编程设计了多种环境——诸如"宴会"、"夜晚"以及"离开"。它使照明环境更加可控，并且由于多数灯具光线较暗，有助于增加灯具的耐用性、降低电力负荷。家庭自动化系统有趣地将光和声对应起来，当业主开启"宴会"照明模式，预设的音乐也会进入播放模式，营造完美的氛围。当然，这些都是根据业主的爱好预设的，但却展现了设计团队为提升业主体验所付出的努力。

# RED Prime Steak

## 融入氖灯的红酒文化
——别克大厦红色牛排餐厅

**Credits**
Client: Red Prime Steakhouse, L. P.
Scale: 557 m², Seating for 150
Architects: Elliott + Associates Architects
Project Team: Rand Elliott, FAIA
　　　　　　　Brian Fitzsimmons, AIA
　　　　　　　Kenneth Fitzsimmons, AIA
　　　　　　　Joseph Williams, AssocAIA
Photography: Scott McDonald, Hedrich Blessing

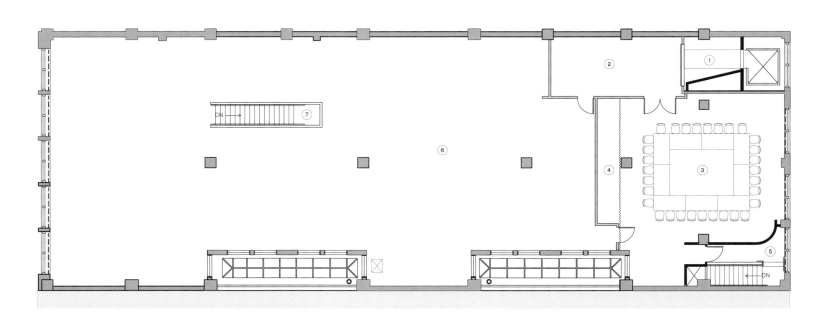

SECOND FLOOR PLAN

1. Elevator Lobby
2. Foyer
3. Banquet Room
4. Storage
5. Stair
6. Banquet Room
7. Stair

The Buick Building is part of the historic downtown district known as "Automobile Alley". This area is a district building on its history as home to more than 50 car dealerships and their related services. Constructed in 1911, the Buick Building was the first automobile showroom in Oklahoma City and began what is now known as Automobile Alley. Unlike most modern sales dealerships, this was a direct sales outlet built and owned by the Buick Motor Company of Detroit. With its limestone ornamentation and embossed name, the two-story structure was well suited to the early automobile industry. Following the April 19, 1995 bombing of the Alfred P. Murrah Federal Building, the development of the Automobile Alley Main Street Program resulted in private investment in excess of $30 million

Thoughts about Red…

1. An experience that awakens / heightens your sense of vision, smell, taste and personal interaction
2. Kitchen as stage
3. Procession
4. Glowing space
5. Comfortable drama
6. Raw and refined
7. Personal scale and grand scale
8. Memorable (at every level)
9. How best to use the Broadway window?
10. "Street magnetism"

The architectural design takes full advantage of 18-foot ceilings, skylights and sheer volume to create spectacular urban beauty and drama. Each table offers an exciting vantage point for memorable dining experiences. The spectacular Wine Wall, some 55 bottles tall by 130 bottles wide, for a total capacity of 7,150 bottles, separates the RED bar from the Main Dining Room. Sweeping the room's grand dimension with a warm glow are suspended "rays" of red neon.

These create energy and light that refract off the building's rustic walls. The concept is called Red Wind.

The rays frame a dramatic procession for diners entering the Main Dining Room. The focal point is the Exhibition Kitchen where red portal highlights the activity and a glowing grill.

RED offers private dining venues – each is a one-of-a-kind setting. Suspended above the Historic Buick's original automobile turntable is a sleek, red lacquer, super-private retreat for memorable events, which seats nine. In the Red Room, slip behind the translucent red fabric panels that separate this private venue from the Main Dining Room. It is secluded, yet the view adds an exciting dimension to the dining experience.

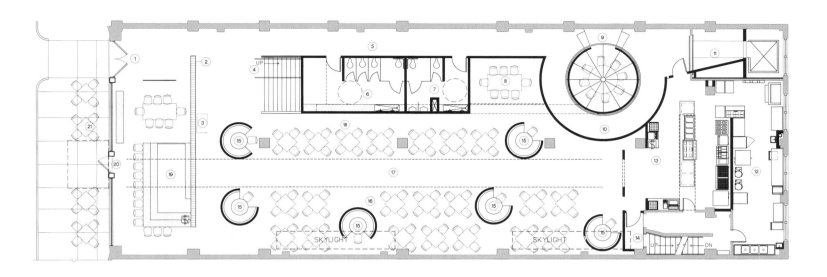

| | | | | | | |
|---|---|---|---|---|---|---|
| ① Main Entry | ④ Stair | ⑦ Men | ⑩ Round Corridor | ⑬ Wait Staff | ⑯ South Dining | ⑲ Bar |
| ② Wine Wall | ⑤ Corridor | ⑧ Steel Room | ⑪ Elevator lobby | ⑭ Stair | ⑰ Procession | ⑳ Historic Entry |
| ③ Rolling Ladder | ⑥ Women | ⑨ Turntable/Red Room | ⑫ Kitchen | ⑮ Tall Tables/Booth | ⑱ North Dining | ㉑ Outdoor Seating |

FIRST FLOOR PLAN

别克大厦是这片著名"汽车街"历史文化社区的一个组成部分。这座大楼以它历史上曾是超过50个汽车代理商和相关服务的总部而闻名。建于1911年的别克大厦是俄克拉荷马城的第一个汽车展示厅，并且成为了现在我们熟知的汽车街。不像大多数现代化的销售代理商，这里是由底特律别克汽车公司建造并持有的直销店。这座两层楼的建筑用它的石灰岩装饰和浮雕名称很好地适应了早期的汽车工业。

关于红色的构想……
1. 唤醒并加强你的视觉、嗅觉、味觉和人际互动的反应
2. 把厨房当成舞台
3. 排列整齐
4. 鲜艳的空间
5. 舒适的戏剧性
6. 粗犷和精致
7. 私人空间和整体空间
8. 每一层都令人过目不忘
9. 使用百老汇窗户有多好？
10. "街道磁性"

建筑设计充分利用了5.5米高的天花板、天窗和绝对的空间营造了壮观的都市美景和戏剧效果。每个桌子都能在用餐过程中提供一个激动人心的优势点。从主宴会厅开始，散布的总共7 150个酒瓶组成了壮观的酒墙，规格为每面55瓶高×130瓶宽。悬浮的红色氖灯为宏伟的空间注入一股股暖流，创造了能量和光线折射出大楼粗犷的墙壁。这个概念叫做"红风"。

灯光排成戏剧性的队列带领用餐者进入主餐厅。展示厨房旁红色的入口突出了室内活动和发光的烤架，是这里的焦点。

"红色"餐厅提供了私人就餐环境，每一个座位都是独一无二的。原来别克的汽车转台被改造成了一个光滑、红漆的超级私人场所，可以坐9个人。在红色房间里，半透明的红色织物从主餐厅分隔出私人空间，既隐蔽，又增加了就餐环境激动人心的维度。

# The Garden Restaurant "CAD" in Kiev

## 隐藏式LED打造的魔幻色彩
——基辅花园餐厅

**Credits**

Location: Kiev, Ukraine
Lighting Design: Ulrike Brandi Licht
Architect: Architektenbüro 4A, Stuttgart
Photography: beproactiv, Moskau

To get to the restaurant the customer passes a wooden tunnel. Indirect lighting emphasizes the warmth of the wood and implements the ease of this place.
Different layers of wooden decks form a landscape in the open yard. A picket fence encloses the artificial landscape and creates an exclusive atmosphere.
The different levels divide the restaurant into several lounges with low level seating furniture.
Hidden LED luminaires create a soft glow and enhance the layer structure of the restaurant. This warm and brilliant lighting brings up an intimate ambiance.
Coloured light in the outlines of the decks add a magical look.

客人将穿过一条木质隧道到达餐厅。间接照明突显了木材所带来的温暖，营造了场所舒适感。
木质台阶层层交叠，在开放院落里形成一处景观。尖桩围栏围合了人造景观，创造出一种独一无二的氛围。不同层级将餐厅分割成多个休息区，每个区都配套低矮的座椅设施。
隐藏式LED灯饰散发出暖色调的灯光，增强了餐厅的分层结构。温暖而灿烂的照明营造出一种亲密的气氛。
彩色灯光框出了台阶的轮廓，更是增添一抹魔幻的色彩。

# Duke of York Square and Headquarters Building

## 白色照明的多样化运用
——伦敦约克公爵广场及其总部大楼

**Credits**
Client: Cadogan Estates
Lighting Design: dpa lighting consultants, Ian Clarke – Senior Designer
Photography: Edmund Sumner

The brief was to create an appropriate night time landmark for this important part of Central London that sympathetically joined the Duke of York Square to the Duke of York Headquarters.

Throughout the design, consideration for local residents and the environment was always significant. The challenge was to create an architecturally informed and powerful lighting response whilst maintaining restraint. Durability and maintainability were also significant considerations.

The Duke of York Square operates as an attractive open space on the Kings Road, one of London's most popular shopping streets. It offers a mix of retail and a fantastic public square with cafes and areas of seating to relax. Although a success by day the owners felt the experience after dark needed to be improved. The requirement was for a vibrant and interesting lighting installation to enliven the space whilst providing a safe evening environment for this city centre location. The key lighting element throughout our solution within the square is the use of Gobo projectors mounted carefully in the trees providing patterned light onto the paving. The pattern simulates sunlight coming through the trees and is deliberately focused to create a soft edge to the pattern. The advantage of removing lighting clutter through this solution was an important part of our philosophy. Further discreet lighting equipment was also integrated into the paving and seating introducing directly illuminated elements at a human scale.

At the Headquarters Building we have created the night time presence by building up the lighting towards the classical entrance portico and providing visual orientation towards the entrances. The building is currently the home of a high profile contemporary

art gallery attracting high visitor numbers from around the world and as such the lighting had to create an appropriate presence. The project utilises carefully controlled recessed asymmetric buried uplights to wash the solid walls between windows at ground floor level. The optics minimise spill light and create a pleasant softness to the buff brick façade. The windows of the main building are internally lit from the bottom upwards, which gives an impression of movement to the ordered fenestration of the façade. A little light is deliberately allowed to spill from the windows catching the horizontal eves detail to identify this important architectural element. Lighting to the entrance portico involves the use of discreet in ground luminaries with appropriate glare control and the required optical control. There was also a minimal use of lighting equipment within the pediment to expose the decorative relief work at high level offering further layers to the lit effect. The surrounding mature trees appear silhouetted against the building as they are up lit on the side facing the gallery façade. This creates a different dynamic from a variety of viewing positions allowing the pedestrian an experience that changes in effect as they move through the square towards the Headquarters Building.

The view from Duke of York Square is significant and the night time connection with the square is harmoniously blended with the pedestrian route to the gallery.

The project is deliberately subtle throughout using warm white light carefully contained, which creates an extremely efficient lighting solution that is architecturally sympathetic and environmentally aware.

设计目标是在伦敦中心城区塑造一处美丽的夜间地标，将约克公爵广场和约克公爵总部大楼连接起来。

在设计过程中，考虑当地居民和周边环境总是非常必要的。而设计面临的挑战在于：需要在静穆的状态下体现建筑的内涵和光照的强度。此外，耐用和维护也是需要考虑的。

约克公爵广场位于伦敦最繁忙的购物街之一——帝王路上，空间宽敞，引人注目。这里汇聚了零售店、华丽的公共咖啡区和休息座位区。虽然白天外观亮丽，但是业主感觉，广场夜间的外观仍需改善。他们要求安装多样化的照明设施，在使城市中心氛围祥和的同时，更平添充满生气的氛围。在规划方案中，广场内核心的照明构件就是图案投射灯，它们被精心安装在树丛之内，向铺石路面投射灯光。灯光模拟成透过树丛的阳光，从而体现出它的柔美感。在设计过程中，去除散光在我们的理念中极为重要。而普通的照明设备则被嵌入铺石路面和座椅上，这套直接照明设施大小与人相同。

在总部大楼处，我们通过增强古老入口大门的光照、入口处的视觉定向来强化大楼的夜间效果。大楼内坐落着高规格的艺术画廊，吸引着世界各地的游客驻足于此。同样，光照设计也需与之相符才行。我们在规划中运用了不对称的壁式上照灯，使光线倾洒在底楼窗户间的墙壁上。

设计师运用镜片消除了溢出光，使浅黄色的砖墙外立面更加柔和。

主大楼的窗户内侧被倒挂式灯座照亮，并使建筑外立面的窗体布局更为动感。而从窗外溢入的一丝光线会照亮夜间的窗体，从而强化这重要的建筑构件。

而入口拱道处的照明设计使用普通的落地灯，并采用眩光控制和光控措施。此外，在山形墙内，最小限度使用光照，最大化地展露装饰画面，从而增加光效的层次感。

当四周长成的树丛面向美术馆一侧的上方被照亮时，其轮廓被背后大楼映衬了出来。当行人从广场向总部大楼走去时，不同的观景点会使人们有着不同的体验，从而营造出不一样的活力。

在约克公爵广场设计中，周围景致也很重要，而广场的夜景与通往美术馆的人行道有机融合了起来。该设计方案通过使用温暖的白色照明设施，使整个规划显得丝丝入扣，无论从建筑还是环保角度都很适合。

# Tokyo-Harajukal Ginza H&M Stores, Osaka, Seoul, Singapore

## 商业照明的典范
——H&M服饰店

**Credits**
Location: H&M Stores – Tokyo, Japan
Client: H&M
Lighting Design: dpa lighting consultants, Richard Bolt, Associate
Interior Design: Universal Design Studios

dpa lighting consultants in association with Universal Design Studios have produced two retail landmark projects in Japan for H&M, the international retail outfit that have entered the Japanese market with the opening of the Ginza and Harajuku stores in Tokyo.

The Tokyo retail scene is innovative and world renowned and hence this was an extremely challenging project, particularly as dpa were competing visually with a lot of high end brand landmark reference points in Tokyo. The idea was to develop building envelopes that would become a landmark installation both from afar and from adjacent with human interaction. The lighting had to ensure suitable impact and dynamics both during the day and particularly at night, which is when the Tokyo retail scene really takes off.

**H&M, Ginza (opened for trading September 2008)**
**UDS Design Concept 1 – Vinyl Graphic – Moiré**
dpa worked in close collaboration with UDS to achieve a stimulating and visually powerful shop front elevation, where the lighting design reinforced the patterning of the glass facade and light box behind. The result of the interplay between artificial

H&M Store, Ginza

H&M Store, Ginza

H&M Store, Ginza

H&M Store, Ginza

H&M Store, Ginza

lighting and applied graphics, is the deliberate visual interference (Moiré) between opposing screen-print patterns, therefore creating a moving dynamic across the three storey shop elevation, which is a very well worked trick. The UDS shop front design incorporates a screen-printed graphic consisting of an opal chevron pattern, which is applied to the internal face of the front window. The front glass is uplit using cool white metal halide luminaires sited in a recessed lighting slot within the shop front floor. Sections of the front glass are also left clear so that a degree of transparency is achieved to view the uplit mannequins, that use warm white metal halide light, and the internally illuminated cool white light box behind. The light box also has an applied screen-print graphic, but uses a black chevron pattern with opal backing, so as to create a distinct contrast with the front glass and maximise the visual impact of the two opposing patterns. The lighting to the light box consists of 5,000K cool white T8 lamps, grouped in to three switching channels, so that there is a degree of flexibility and control over the intensity of the shop front lighting, which is controlled via a solar time clock.

Lighting trials were conducted in Tokyo and also in Stockholm (home of H&M) utilising an existing H&M shop front. It was really important that the selection of the lamp colour temperature associated with the luminaires were carefully considered and trialled against the proposed materials and graphic patterns, so that we could be confident of a stunning result. In addition, the trials would be used to determine the required intensity of light to appropriately express the façade and reinforce the drama of the design, taking in to consideration the active and energetically lit Tokyo retail street scene.

H&M Store, Harajuku

The concept was also applied to H&M Shinjuku, Japan.

**H&M, Harajuku (opened for trading November 2008)**
**UDS Design Concept 2 – Matt White Slats**

H&M required both the Ginza and Harajuku projects to provide a visually stimulating lit result at night, however the interior design concepts for both were quite different, together with the H&M buildings in which the stores were housed.

Harajuku differs architecturally from Ginza in that the façade verticality, appears to be made up from a series of overlapping glazed cubes, whereas the Ginza building has a somewhat more regular linear form, but with a central sinuous façade line that neatly delineates the solid and glass cladding layout, a feature which is particularly strong on the lower shop floor levels. At night the Harajuku building appears to resemble a collection of lit "ice cubes" that are built up, in to an architectural stack, however the main attraction is accommodated within the lower three retail floor levels. Here, the glass façade encases a clever arrangement of matt white painted aluminium slats, which lean forwards and backwards, resulting in a wave like pattern. The lighting concept was to express the form of the slats on both sides of the façade and work with their orientation, so that a sense of movement could be achieved. In addition, dpa were keen to work on the interplay of light and shadow, in relation to the light travelling through the gaps between the slats. We wanted it to appear as through it was natural light that was creating a pattern on the surface beyond the slats for added theatre.

Following a series of lighting trails in Tokyo, the final scheme allowed for the integration of uplights within the shop front floor, using cool white 4,200K metal halide lamps and the use of a spreader lens optic, which produced a linear wash of light. The lighting effects were reversed for the internal face of the slats, which were down lit via track mounted spotlights, again with spreader lenses. Therefore, a pattern of lit lines is cast both on to the shop front floor and also the ceiling of the store.

UDS transferred their interior concept to the central staircase, which was lit via recessed downlights and track mounted spotlights with spreader lenses. The staircase is something that you might expect to find in an art gallery or museum, with its cleverly crafted sculptural form.

The concept was also applied to H&M Osaka, Japan and to the staircase at H&M Seoul, South Korea.

H&M Store, Harajuku

H&M Store, Harajuku

H&M Store, Singapore

dpa照明咨询公司与通用设计研究室精诚合作，在日本建造了两栋地标性的H&M零售商场。作为世界级的零售商，H&M在东京银座和原宿开设了商场，以此宣告进军日本市场。

东京的零售业日新月异，举世闻名。因此对于dpa咨询公司而言，这是一个极富挑战性的项目，尤其是他们要与东京的大量高端品牌地标参考点竞争。设计方想塑造一处大楼外墙，使其成为远近皆知的地标设施。需要全天候保持活力和效果的照明，尤其在夜间，东京的零售街正是人头攒动的时候。

**银座的H&M服装店（2008年9月开业）**
**通用设计研究室方案1——乙烯图案——叠纹**

dpa照明咨询公司与通用设计研究室通力合作，使商店正面外观激动人心且极具可看性，那里的照明设计强化了正面玻璃和后方灯箱的形态。人工照明和规制图形交相辉映，形成了反向丝印图案的叠纹，由此使三层楼的商店营造出动态效果，实是巧夺天工之举。

通用设计研究室将店门前沿设计成一幅丝印图案，图案由锯齿形花纹的玻璃构成，该结构被运用于前窗内侧。前门玻璃被下方冷色调的金属卤素灯照亮，灯具安装在商店前门地板的嵌壁灯槽内。商店前门玻璃也被照亮，以便使人清晰地看见店内的人型模特造型，而透明的效果也源于金属卤素灯以及后方用于内部照明的白色灯箱。灯箱设计也采用了乳白衬底的黑色丝印图案，从而塑造前门玻璃的对比效果，并与两种图案的视觉效果相得益彰。灯箱的照明设施为5 000度色温的T8灯管，通过三个通道进行切换，以便增加灯光的多变性，并依靠太阳能时钟对商店正门照明强度进行控制。

H&M公司在东京和斯德哥尔摩本部，利用公司商店的前门进行了光照测试。其中仔细考虑灯光色温和灯具选择，并对选定的材料和图案进行测试是极其重要的，如此便可对结果了然于胸。此外，通过测试还可以发现，何种程度的光照能够烘托建筑正面外观，并强化设计的戏剧效果，使东京零售街道在光照下更为活力四射。

该理念也被运用于H&M位于日本新宿分店的设计。

**原宿的H&M服装店（2008年11月开业）**
**通用设计研究室方案2——哑光白色百叶条**

在银座店和原宿店的两个项目中，H&M公司都想着力营造出完美的夜景效果，然而内部设计的理念和商场建筑二者有很大区别。原宿店和银座店建筑角度的差异在于商店正门的垂直度。原宿店主要依赖一组重叠的玻璃方体营造垂直效果，而银座店大楼则呈常规的线形，但大楼中央曲折的外观整洁地描绘出固态的玻璃层，这样的外观在底层商店尤为明显。夜晚原宿店商场内部如同竖立着几座闪耀的"冰体"一般，这些"冰体"已经融入建筑的夹层中。然而最引人注目的还是大楼的下三层。在那里，建筑的玻璃外墙将哑光白色铝制百叶条包裹在内，百叶条前后摆动，营造出波浪状纹理。

这里的照明设计是为了烘托外立面两侧的百叶

# Bi-Tan Bridge

## 光与桥的对话——碧潭吊桥

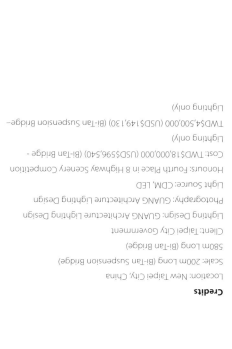

**Credits**

Location: New Taipei City, China
Scale: 200m Long (Bi-Tan Suspension Bridge)
580m Long (Bi-Tan Bridge)
Client: Taipei City Government
Lighting Design: GUANG Architecture Lighting Design
Photography: GUANG Architecture Lighting Design
Light Source: CDM, LED
Honours: Fourth Place in 8 Highway Scenery Competition
Cost: TWD$18,000,000 (USD$596,540) (Bi-Tan Bridge - Lighting only)
TWD$4,500,000 (USD$149,130) (Bi-Tan Suspension Bridge—Lighting only)

Bi-Tan Suspension Bridge, one of famous sceneries in Taiwan, is a cable suspension type and pedestrian bridge. Bi-Tan is the name of lake which the bridge goes across. As the name it is called, Bi literally means Green and Tan means Lake in Chinese. The bridge is approximately 200m long which is a connecting bridge east and west side of the lake, 3.5m in width, and the tower is 20m in height. Its construction was finished in early 1937 and has become a symbolic landmark in Xindian District, New Taipei City. The bridge was renovated by local government in 2000 and planned lighting design in 2007.
Bi-Tan is a historical region. As time goes by, local businesses faltered and had been replaced by new neighbour regions. Thus the lighting scheme performs a sense of region rebirth through the light in Bi-Tan. The plan influenced the residents' living environment and improved local business activities, also changed the image of city view. The concept focused on creating a communication between two bridges: Bi-Tan Suspension Bridge and Bi-Tan Bridge.

**Pearls in the Sky**

The idea behinds the design is stringing lights as a string of pearls in the sky. 184 5W RGB LED Pointolite lights are attached on major arch cable and randomly change colours to emphasize bridge's elegant features. Two continuous linear lights, which are consisted of 370 LED linear luminaires, are against outside of two rails to floodlight rails with 3000K colour temperature from down to top. Moreover,

■ —— Current 3
"会呼吸"的互动性灯光设计

When at rest and no pedestrians are on the bridge, Current 3 "breathes". With a single person crossing the bridge, the effects follow them before returning to the breathing mode.

休息时间，桥上无人时，Current 3会发出"自然呼吸"。一有人经过桥上，灯光会在整个过桥过程中，会跟行人的步伐变化而变化。

■ —— The Rion Antrition Bridge in Greece
希腊雷南·安托雷南大桥

The illumination of the deck was intentionally designed to produce a super positioning of shadows bringing relief, texture and life to the immense length of the bridge.

桥面的照明设计在于刻意产生大尺度的影子，浮雕效果、质地和生命力被带入这座极度长的大桥。

■ —— Telekom Bridge
电信步行桥

This experience of a physical "light-shadow" following the person gains yet another dimension when several pedestrians linger in the vicinity of the tower.

这是一种多个个人的"光影"体验。当越来越多的行人因电梯塔而滞留时形成一种独特的感觉。

■ —— Prins Claus Bridge
克劳斯王子桥

Immense care was taken in all aspects of the design process to reduce energy consumption and seamlessly integrate the new lighting with the Bridge.

在设计过程中，设计方注重降低能源消耗，完美融合新方案的新型照明用具与桥梁。

■ —— Bi-Tan Bridge
碧潭吊桥

The colourful light scheme and flicking lighting movement bring the bridges alive. These ideas influenced on region's night image and made an outstanding landmark to the city.

缤纷的灯光明暗及色温的变化，呼应水的流动，远眺桥点亮了大桥优美的山峦起伏上发亮球，点亮串起，随着灯光脉动呈现浪漫的夜景意象。

# 03

## Transportation
## 交通

H&M Store, Seoul

H&M Store, Singapore

来,并与其他货架的一起刻造动感效果。此外,dpa还加入可识别其他楼层的独特的照明设备,使来访者以及对实际销售路线进入店内,就好像在其中蜿蜒曲折地穿梭的松鼠自然发光。

在未来几年内,这一系列发展规划,最终将运营分店的门类向上向外拓展扩大,使用总面积为4,200平方米的图案装扮亮灯,并借助了聚光灯效果,将沉出该性质的光效。这些设计与周围的产品、商品用特别直竖装饰灯,并让灯光的穿透力,同样地表达了聚光灯效意图。因此,有店内的员工注意到他们的表情可以紧紧吸引眼光的存在,是因为设计师把客体他们的表情为这发现引入了中心悬挂,那些玻璃镜至周围如此变换的蓝光出奇。同意只有只有在美术作品的摆放设置,才有各种加工艺精细的搭配标志诸真。

该建设也被用于H&M日本大阪分店的独特国具分分的领先榜样。

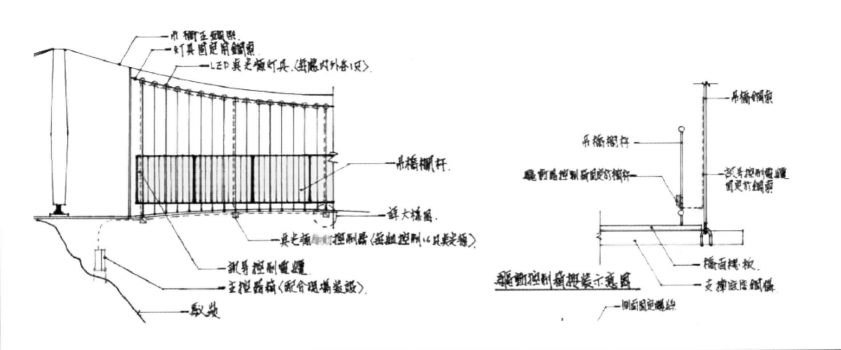

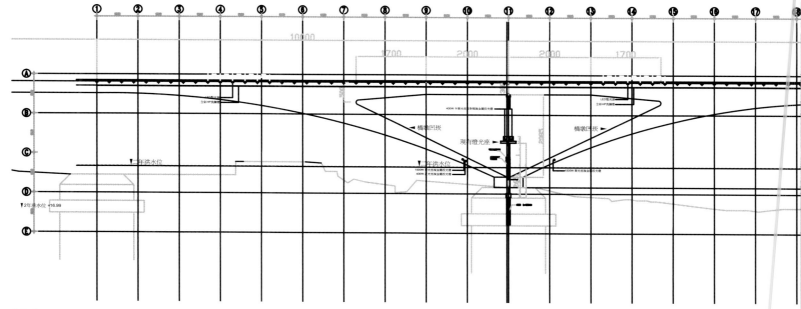

customized lighting fixtures are evenly attached on rail frame's vertical materials for indicating the pathway and floodlighting bridge's deck. Flashing pointolite lights and static floodlights are reflecting on flowing water to compose a harmony night environment in Bi-Tan.

**Communications between Bridges**

Bi-Tan Bridge is located at National Freeway No.3 and beside Bi-Tan Suspension Bridge. It was completed in 1997 and the lighting design was planned in 2008. Different to suspension bridge, Bi-Tan Bridge is concrete Box Girder Bridge which has two stylish V-piers cross water and 850m long. The light scheme is planned with 16 1,000W CDM projecting lights and 8 400W CDM floodlighting for highlighting the feature of piers. The luminaires were fixed at piers' façade and aiming opposite directions to give emphasis to its V characteristic. The ideas for lighting up the bridge deck, 600 RGB wall-wash luminaires were concealed at two sides of bridge deck and washed concrete façade with changed colours. In order to link the idea of two bridges, the design applied 242 LED Pointolite luminaires at the edge top of Bi-Tan Bridge to respond to the design spirit of Bi-Tan Suspension Bridge which is a string of flicking pearls in the sky.

The colourful light scheme and flicking lighting movement bring the bridges alive. These ideas influenced on region's night image and made an outstanding landmark to the city. Thinking about how people interact with the bridges and understand the history of region, lighting created the communications between two representative bridges, and people got the opportunities to get into the communications by viewing the lighting performance. After the lighting scheme of region rebirth, the place will keep remain the Bi-Tan's history for the future.

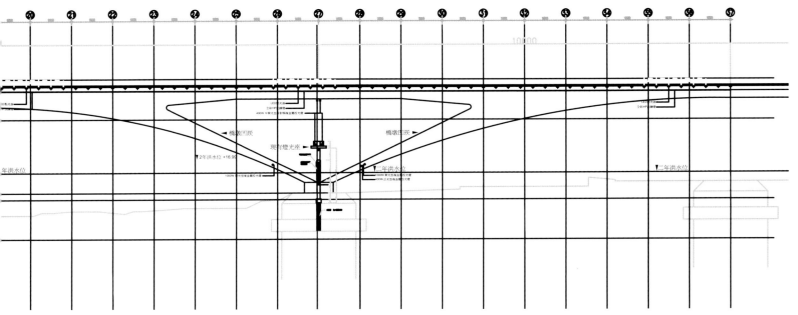

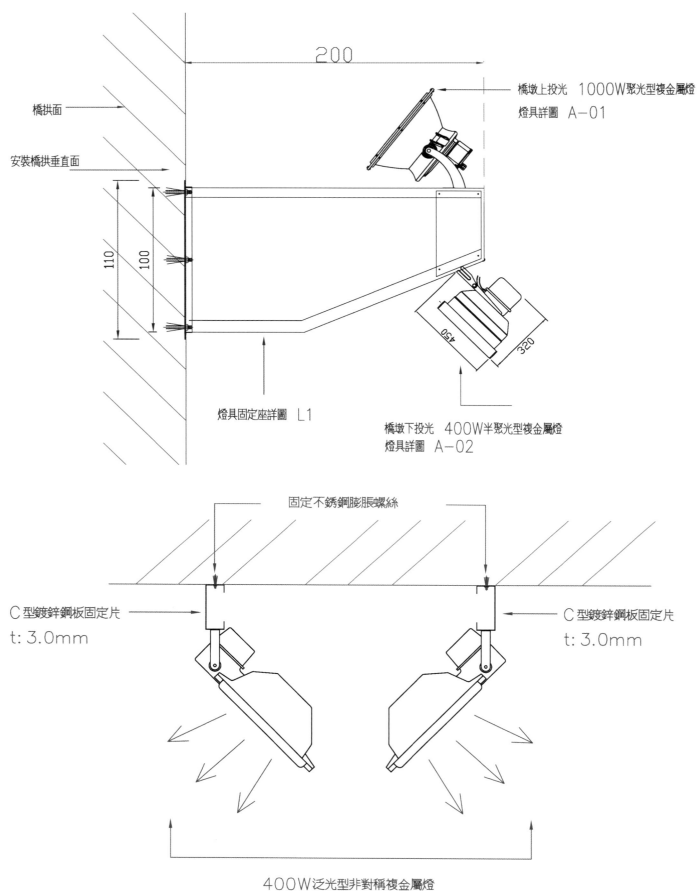

坐落于台湾新北市新店的碧潭吊桥，自1937年启用至2000年依原貌重建，为钢索桥，长200米，宽3.5米，高20米，由早期的交通辅助路线，渐渐转变成休憩游乐场所。碧潭，如其名拥有碧绿色的湖面及两岸高耸山壁围绕，由新店溪汇集而成，形成一种城市中与世隔绝的天然景色。

碧潭的历史悠久，随着斗转星移，地区没落，因此该规划设计是希望藉由照明的改变带给区域崭新的重生感。其中主要概念以吊桥及相对望的北二高碧潭大桥相互呼应为主题，为吊桥增加夜间灯光装饰及透过北二高碧潭大桥上LED的光彩流动，让两桥相互对话，形成了过往所不曾出现的璀璨景致，随着光线律动，为新北市增添一座新的艺术地标。

**如珍珠般璀璨，颗颗串起**

碧潭吊桥以184颗5瓦的RGB全彩LED点光源点缀于吊桥主桥缆，如串联的璀璨珍珠高挂于天际勾画出缆绳吊桥的特殊优美弧度。两侧扶手外侧布满共370盏的线型LED灯，以色温3 000K的光附着于栏杆上。订制的灯饰规律地吊挂在栏杆内侧，泛光布满走道，并于黑夜中为行人指引出明亮的道路。夜间点光源以缓慢的频率闪烁变化，稳重固定的暖白洗光，倒映于湖面的光影流动，打造出碧潭整体环境的新生气息。

**桥与桥的对话**

不同于碧潭吊桥丰富的历史形象，横跨新店溪的北二高碧潭大桥长约850米，是国内目前最长的预力混凝土箱型桥梁，于1997年完工，新北市(旧称台北县）政府在2008年的大碧潭旗舰计划即包括了北二高碧潭大桥照明改良的施作，以增添碧潭的美丽夜色。桥体结构分解为两座V型基础座横跨碧潭湖面，桥体则为箱型混凝土架构于基础座上。将16盏1 000瓦及8盏400瓦的复金属投、泛光灯以上下投泛光的形式固定于基础座立面，随着光线布光的角度，强调山桥墩的V型弧度。箱体立面则布满600盏RGB全彩LED线型洗墙灯，随着LED色彩温度的变化，改变了混凝土的原色。桥身最上层则为242颗RGB全彩LED点光源，与碧潭吊桥上的点光源相互辉映。

藉由照明色彩及色温的变化，呼应水的流动。远眺点点灯火恰似文山堡颈上珍珠，颗颗串起，随着光影时而迷离时而缤纷。二桥的对话，以灯光的变化流动相互呼应。照明改变了碧潭的夜晚，让景色变得丰富生动，更多的改变不仅仅是夜间的形象，同时也活络了当地的生活形态及文化。

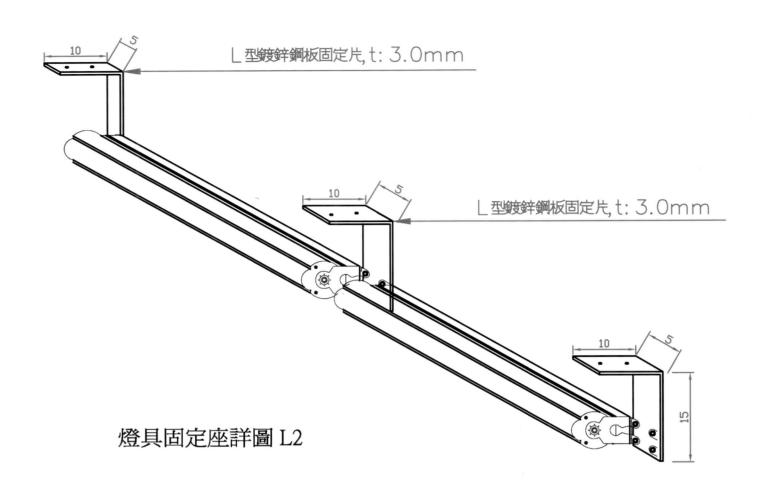

燈具固定座詳圖 L2

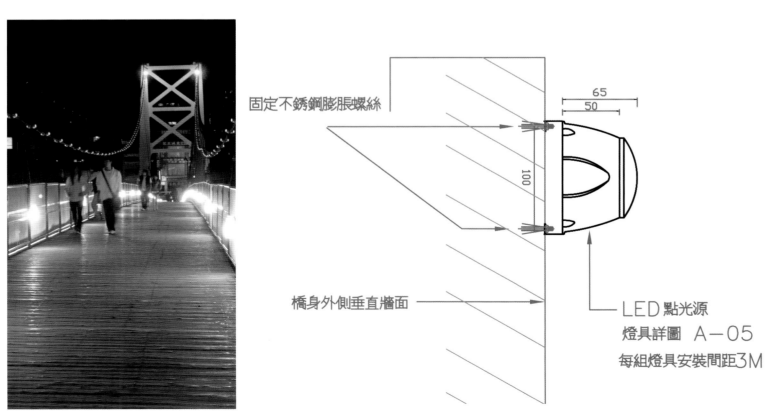

LED點光源
燈具詳圖 A-05
每組燈具安裝間距3M

# Prins Claus Bridge, Utrecht, The Netherlands

## 低能耗照明
——克劳斯王子桥

**Credits**
Lighting Design: dpa lighting consultants, Nick Hoggett, Partner
Architect: UNStudios
Photography: Wim van Ijzendoorn

Prins Claus Bridge was completed in 2003 but originally without any architectural lighting due to budget constraints at that time. The Architects for the bridge UN Studios are renowned Amsterdam based practice had always intended for the project to be sensitively illuminated. The structure quickly became an icon for Utrecht during the day linking the city to the new and important commercial district of Papendorp. In 2009 the city recognised the value of sympathetically illuminating this elegant feet of engineering, with its sculptural qualities so the Bridge could be fully enjoyed at night.

dpa lighting consultants worked closely with UNStudios and the Client to design a lighting solution that emphasised the sculptural qualities of the structure, whilst being careful with energy consumption. The idea to focus attention on the main pylon with its complex shapes, and the points at which the bridge connects to the ground either side of the canal quickly materialised. The lighting has been designed to show off the shape of the Pylon, with one face not directly lit, other surfaces illuminated and shadows from the cables allowed to immerge.

The bridge appears differently as you move around the adjacent areas, and particular effort was put into not illuminating the full lengths of the cables, and just allowing a brush of light fanning across them as they get closer to the structural pylon. This was carefully considered both during the design stage as

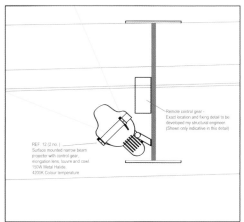

Section Detail E-E / through central spine showing position of dpa reference 12   1:10@A1

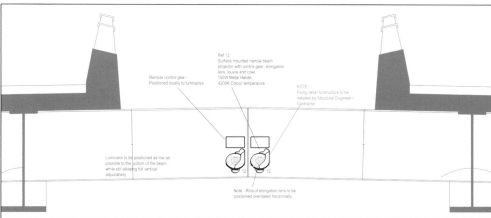

Section Detail F-F / through central spine showing position of dpa reference 12   1:20@A1

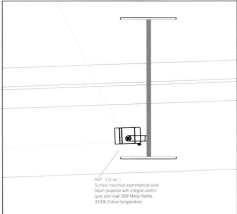

Section Detail G-G / through central spine showing position of dpa reference 3   1:10@A1

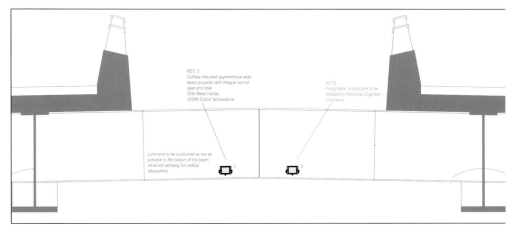

Section Detail H-H / through central spine showing position of dpa reference 3   1:20@A1

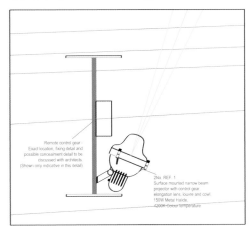

Section Detail A-A / through central spine showing position of dpa reference 1   1:10@A1

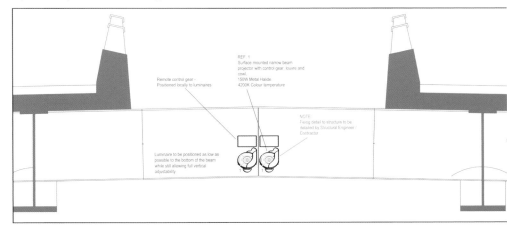

Section Detail B-B / through central spine showing position of dpa reference 1   1:20@A1

a desktop study using carefully controlled projectors and the results were verified during full scale lighting trials at the site.

The face of the Pylon orientated towards the city, is illuminated with carefully controlled narrow beam metal halide projectors, some with lenses located in the aperture between the two road decks so that they cannot be seen during the day, and do not glare drivers at night. The two side surfaces facing the canal are illuminated from again carefully controlled metal halide projectors, on a specially designed column, and the other surface is not illuminated at low level with just some narrow beam metal halide projectors, picking up the top of the pylon and creating shadows from the cables.

As stated above full scale site trials took place to provide the client with an actual real life view of the proposed lighting solutions. The advantage of having an existing structure is that such trails can take place to inform the client to take away ambiguity of communication. The trials also allow the design team to refine the proposal carefully considering every shadow and detail.

Below the road deck the main pylon connects to the

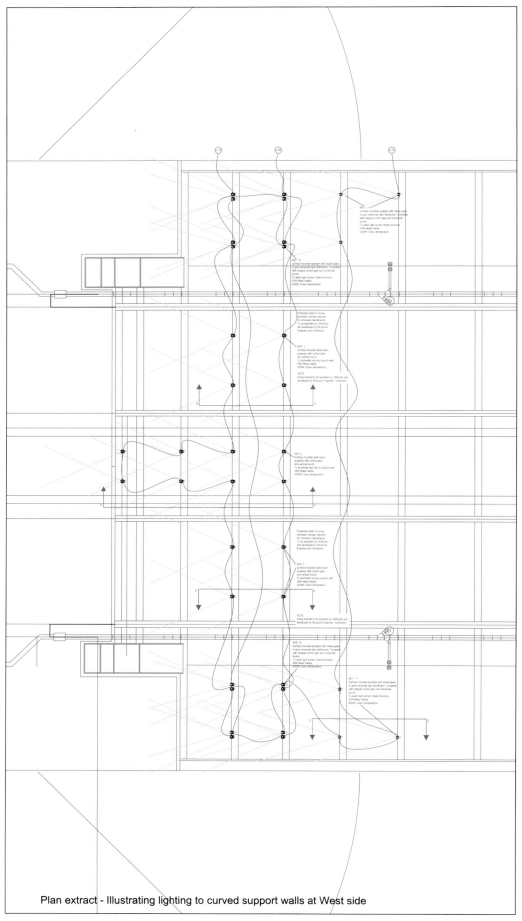

Plan extract - Illustrating lighting to curved support walls at West side

surface in a shape like a paperclip. Metal halide projectors have been located below the road to reinforce this shape visually, and connect the pylon with the ground. Again, as with the pylon above the deck, some surfaces are lit, and some are not to provide interesting modelling and emphasis to the elegant but complex forms.

The underside of the road deck connects with the ground with a sweeping curve, and these curves are illuminated to provide a visual connection of the road deck to the banks of the canal. Efficient metal halide projectors carefully controlled have, as with other elements, been utilised here too. Immense care was taken in all aspects of the design process to reduce energy consumption and seamlessly integrate the new lighting with the Bridge.

The lighting equipment was supplied by Sill Lighting and Meyer Lighting, both German manufactures with Osram metal halide lamps.

The lighting trials explored various techniques to expose the structure and this inspired us to create two different lighting states. The reason for creating two lighting scenarios is to provide two different night time presentations of the structure. One lighting state consists of all the lighting elements turned on and this is instigated at dusk, staying on until about 10 pm. The second lighting scenario turns off the lighting to the side of the pylon above ground and with this lighting off leaving the majority of the sides not directly illuminated a different sculptural vision of the structure is created. This play with light is in keeping with the complexity yet visual simplicity of the bridge and adds another dimension for the viewer. It also further saves energy consumption after 10 pm, which is at a time when there is less traffic in this area.

The project was very special to dpa and treated like a piece of art, the results of which everyone concerned is proud of, and we hope locals in Utrecht and visitors all enjoy the night time view of this special canal crossing.

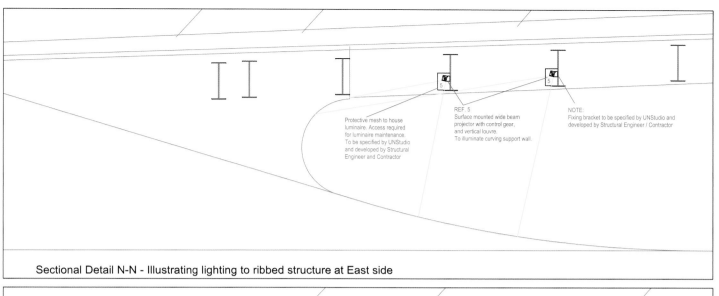

Sectional Detail N-N - Illustrating lighting to ribbed structure at East side

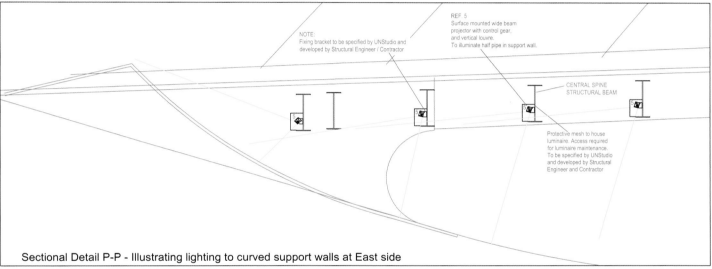

Sectional Detail P-P - Illustrating lighting to curved support walls at East side

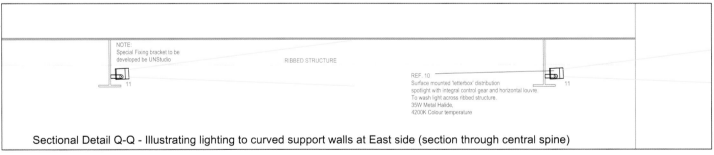

Sectional Detail Q-Q - Illustrating lighting to curved support walls at East side (section through central spine)

Sectional Detail R-R - Illustrating lighting to curved support walls at East side (section through central spine)

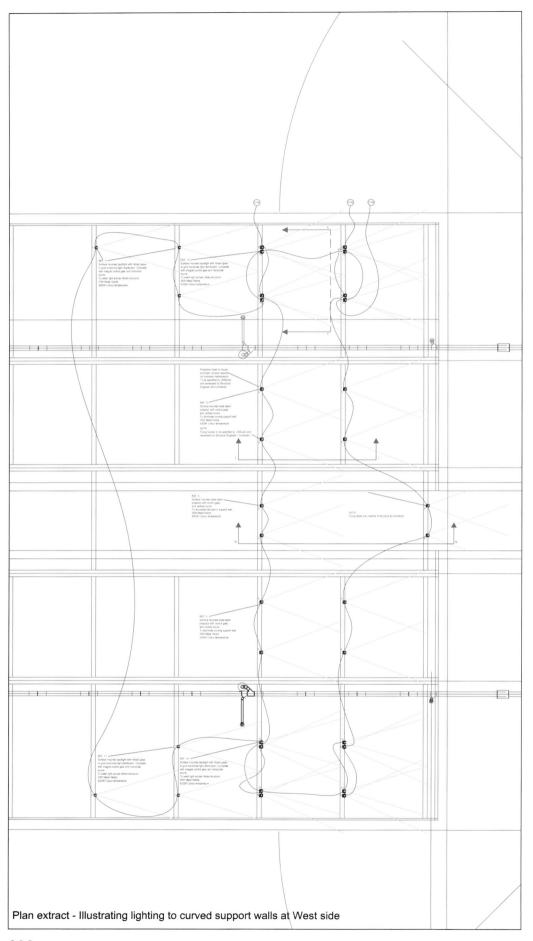

Plan extract - Illustrating lighting to curved support walls at West side

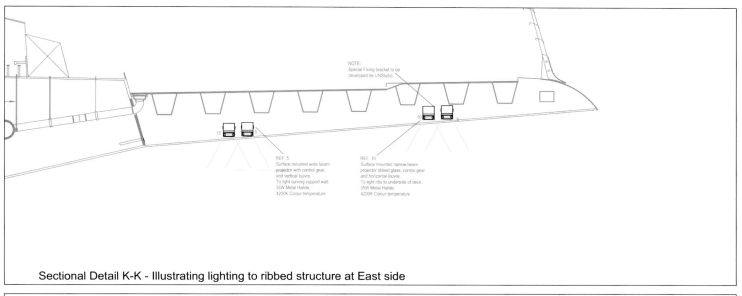

Sectional Detail K-K - Illustrating lighting to ribbed structure at East side

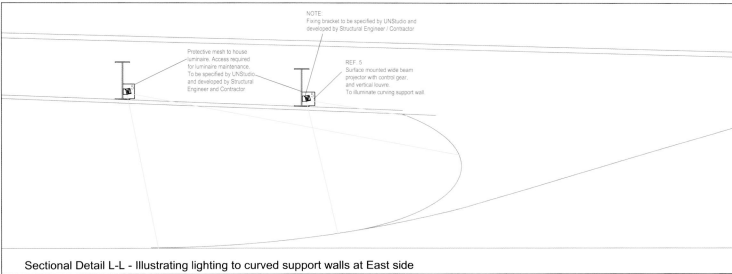

Sectional Detail L-L - Illustrating lighting to curved support walls at East side

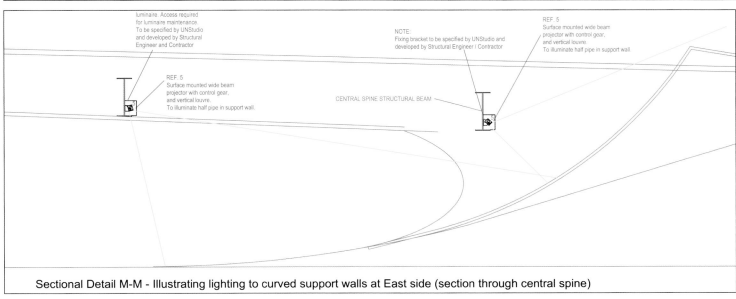

Sectional Detail M-M - Illustrating lighting to curved support walls at East side (section through central spine)

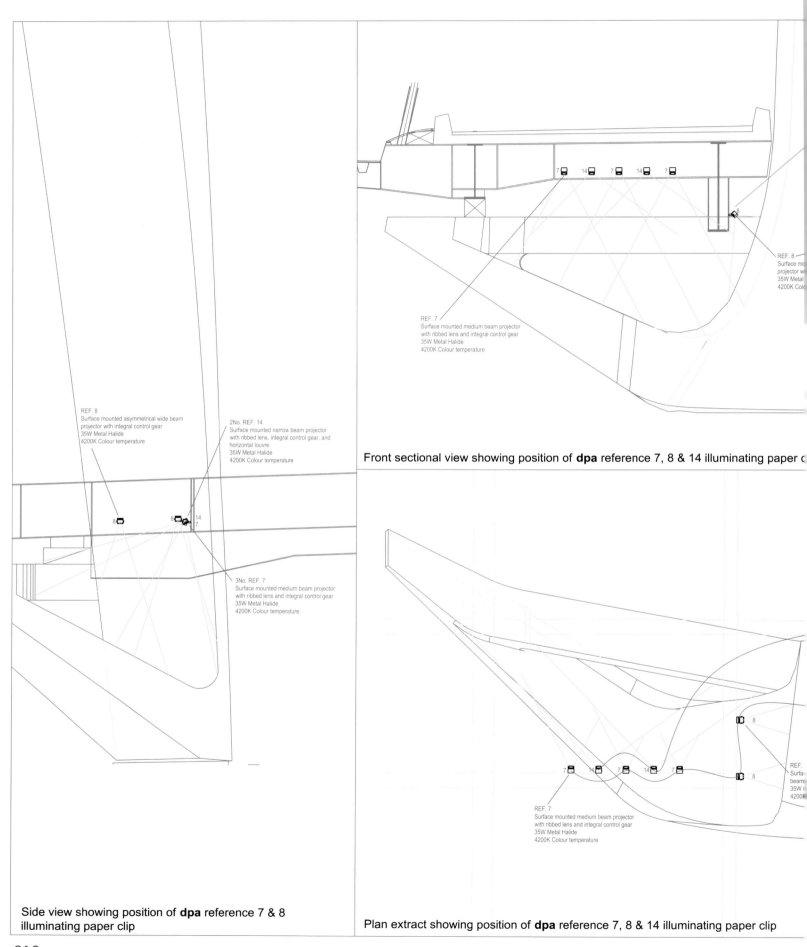

REF. 14
Surface mounted narrow beam projector
with ribbed lens, integral control gear, and
horizontal louvre
35W Metal Halide
4200K Colour temperature

REF. 14
Surface mounted narrow beam projector
with ribbed lens, integral control gear, and
horizontal louvre
35W Metal Halide
4200K Colour temperature

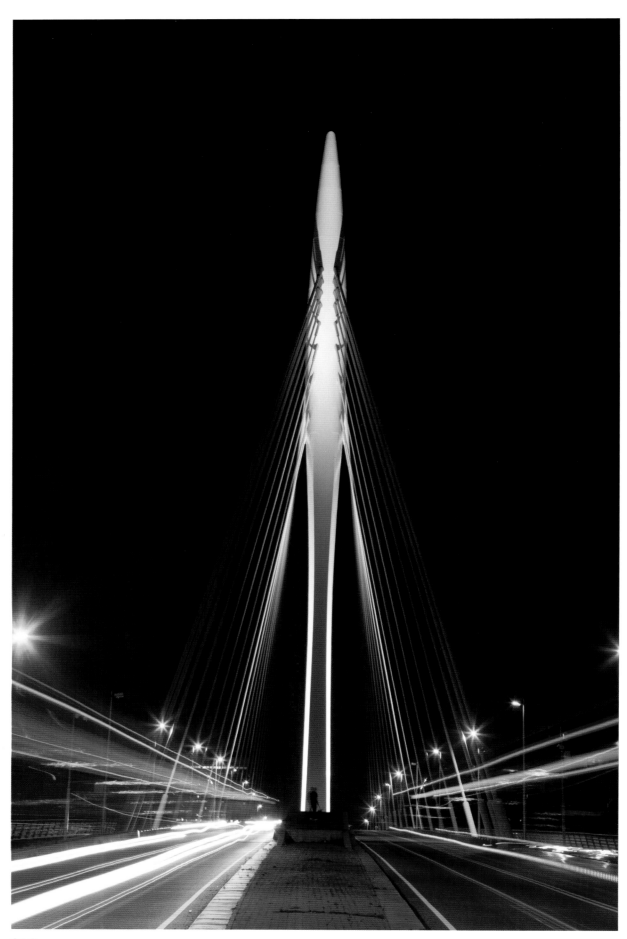

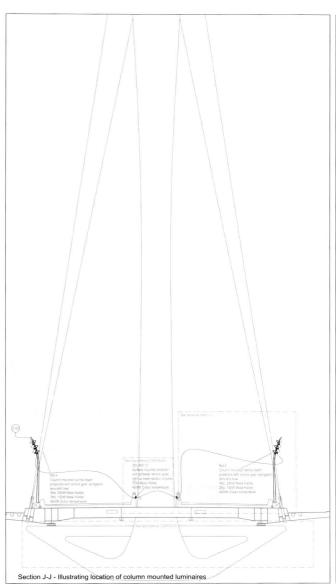

Section J-J - Illustrating location of column mounted luminaires

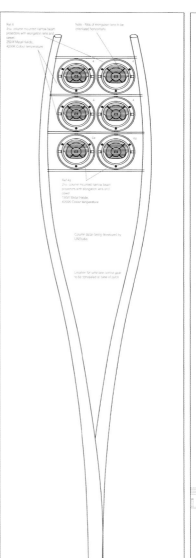

Front elevation of product

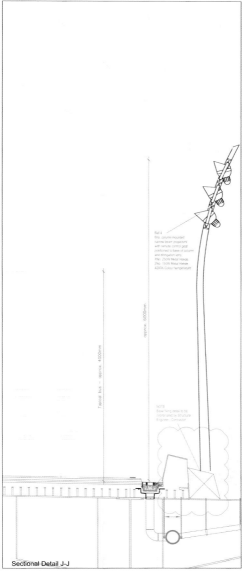

Sectional Detail J-J

克劳斯王子桥建成于2003年，但由于当时预算紧张，未能进行照明设计。负责该桥梁设计的公司——UNStudio设计事务所来自阿姆斯特丹，他们始终致力于光照敏感的建筑规划。自从桥梁将城区与新建的帕潘多普重要商业区连接后，桥梁便成为乌德勒支当地地标。2009年，市委会认识到适宜的光照可以使桥梁工艺得以展现，使其在夜间也可以展现其雕塑般的形态。

dpa照明咨询公司与UNStudio设计事务所及客户们合作设计出一套照明方案，能在保证能效的前提下烘托桥梁结构。以造型复杂的大桥指示台、桥梁与运河两侧地面的交界处为中心点的想法很快形成。设计的照明需烘托桥塔，使其三面照光，一面间接透光，阴影处以电线遮挡。

当您走到桥对面时，桥的外观开始变得不同。此时不应追求对整座电缆进行照明，而将光线倾洒在电缆靠近桥塔的地方。这个问题不仅在探照灯使用研究时，还在全方位灯光测试核实效果期间进行考虑。

桥塔面朝城市的一侧被来自金属卤素灯狭窄的光柱照耀着，部分灯具的支点位于两处道路面层间，白天无法被察觉，且在夜间不会影响司机的视线。桥塔面朝运河的两侧同样被两处特别设计的金属卤素灯照亮。桥塔剩余的一侧位于低处，金属卤素灯狭窄的光线无法触及，因为灯具仅能照射到塔顶，却无法穿过电缆进行照明。

至于上面提及的全方位灯光测试，那是为了让客户亲自体验设计的灯效。保留原先架构可使客户摒弃陌生感。测试的目的也是为了让设计团队从各个细节中对规划进行改进。

位于道路面层下的主桥塔连接一处形似回形针的地带。安装在道路下方的金卤探照灯更加烘托了这条路的形状，同时连接桥塔和地面。至于道路上方的桥塔，部分塔身受光，而另一部分高雅、复杂的外观并未得到强化和烘托。

道路面层的下侧连接曲线形的地板，被照亮的曲线轮廓使道路外观与运河河岸互相连接。高效的金卤探照灯和其他构件也在此得以运用。

在设计过程中，设计方注重降低能源消耗，完美融合新安装的照明用具与桥梁。

照明设备由"希尔照明"与"梅耶尔照明"厂提供，它们是来自德国的制造厂，专门生产欧司朗

金属卤素灯。

通过光照测试，设计方探求出了烘托建筑架构的数个方法，这激发了我们营造两种照明状态的灵感。之所以制定该方案，是为了营造两处不同的桥梁夜景。一种照明状态便是：将所有照明设施在黄昏时分打开，持续到晚上10点。第二种照明状态是：关闭朝向地面上桥塔的光照，随着灯光关闭，使大部分地方无法接受直接照明，却恰好烘托桥梁的建筑结构。如此的灯光效果使桥梁外观既简约又复杂，使欣赏者看到一副别样场景。且由于该地区此时人烟稀少，所以10点之后照样可以实现能耗的节约。

该项目对dpa公司意义非凡，如同一幅艺术品那样受人关注，这也是我们的骄傲。我们希望乌德勒支当地居民和游客能齐聚运河河畔，享受这美丽的夜景。

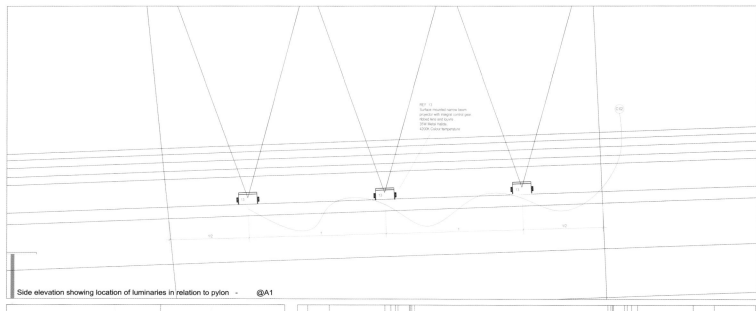

Side elevation showing location of luminaries in relation to pylon - @A1

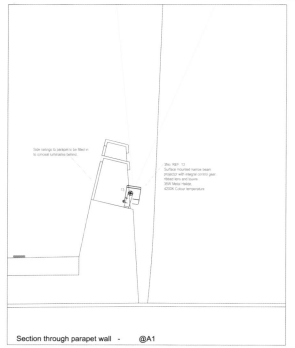

Section through parapet wall - @A1

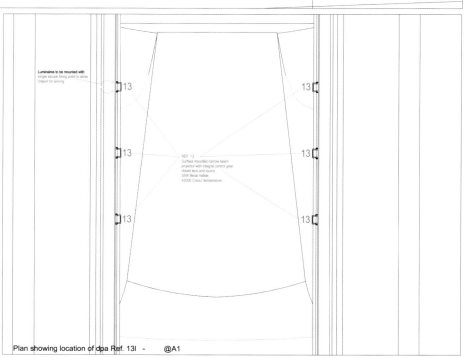

Plan showing location of dpa Ref. 13l - @A1

# Telekom Bridge

物理"光影"体验
——电信步行桥

**Credits**
Location: Bonn, Germany
Scale: 270 m²
Client: Deutsche Telekom AG
Lighting Design: Licht Kunst Licht AG
Lighting Designers: Florian Amannt, Dipl. Dipl.-Ing. Stephan Thiele, Dipl.-Des.Thomas Möritz, Prof. Dipl.-Ing. Andreas Schulz
Architects: Schlaich Bergermann und Partner Structural Consulting Engineers, Stuttgart, Germany
Photography: Lukas Roth
Installation Cost: 580 000 Euro
Watts per sq. ft. (or meter): 18W/m²

### A Connecting Experience—The Telekom Bridge in Bonn

Fast connections are their specialty—as Europe's largest telecommunications company Deutsche Telekom AG operates all networks so that man and technology can communicate via cable, satellite and radio. Sometimes, however, we prefer to talk face to face—something we all have in common. For many employees of the Telekom headquarters in Bonn this has been made considerably easier now that a new footbridge connects two office blocks across a busy road. A lighting concept with interactive elements orchestrates the sweeping filigree architecture of the bridge and follows the client's company slogan: staff members and the public can now "Experience the Connection" on the bridge.

The 72-metre bridge is as much a boon for the 4,000 Telekom employees working on either side of the B9 federal highway as it is a gift from the company to the inhabitants of Bonn. Before the bridge's construction pedestrians crossing the Friedrich Ebert Allee (B9) had a 600-metre journey—a distance that many attempted to short-cut by hazardous sprints across four lanes of roadway and two tram tracks. Safety was thus the pivotal motive for bridging the road at 7 metres above ground. Comfort was a second motive and consequently less mobile people now have elevators at their disposal on both sides of the bridge. The third intention was to create a piece of corporate identity. With the sweeping deck of the bridge supported by just 5 slender steel tubular columns the engineering office of Schlaich Bergermann and Partner have designed a structure that conveys modernity and technical mastery. A particularly clear reference to Telekom AG is made by the lighting concept, since it allows the client recourse to its CI colour.

### Video Pixels in Magenta

A magenta-coloured luminous band spans the Friedrich Ebert Allee day and night. On both longitudinal flanks a 65-metre long media display has been integrated into the bridge construction. These LED video display panels have a resolution of 32 x 6,500 pixels and a colour saturation of 16 bit. They consist of a seamless array of display modules which were especially developed for this project.

From the outset the media strips were incorporated into the architects' design process. As a result the LED modules and their control electronics, mains units and cabling are coherently integrated into the bridge. When switched off, the media panels do not betray the fact that they are idle video screens but appear as a self-evident part of the structure. All video displays and system components are easy to maintain and

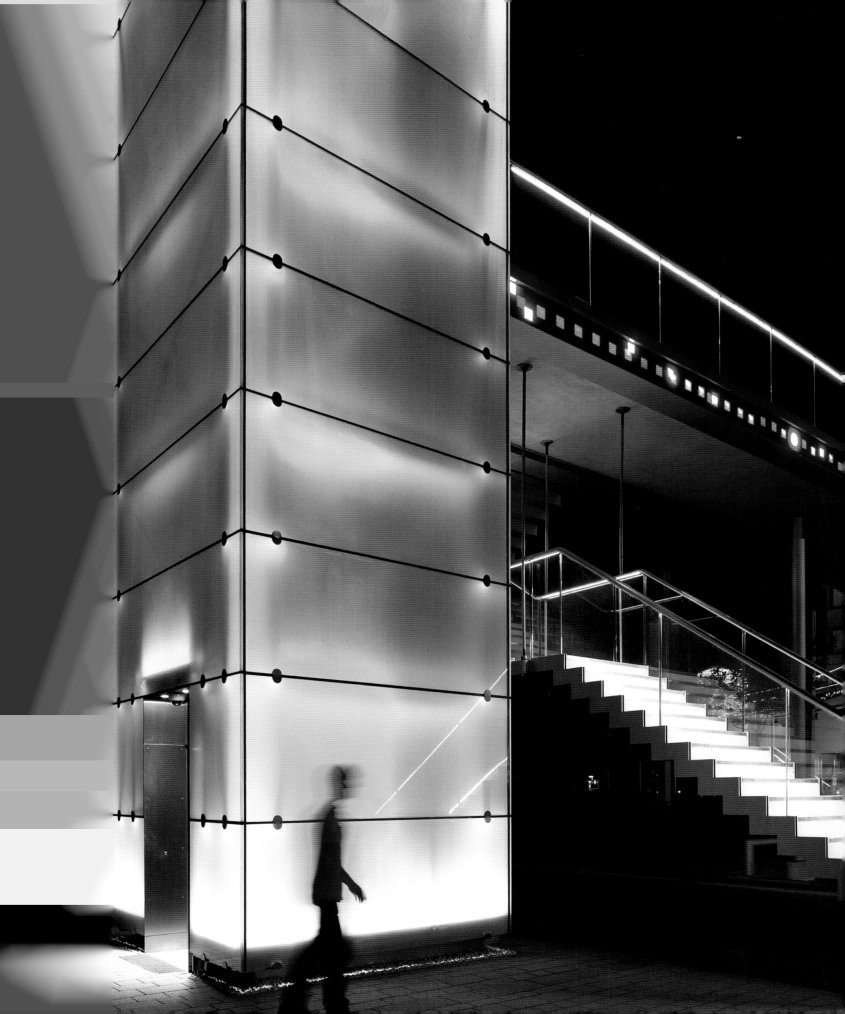

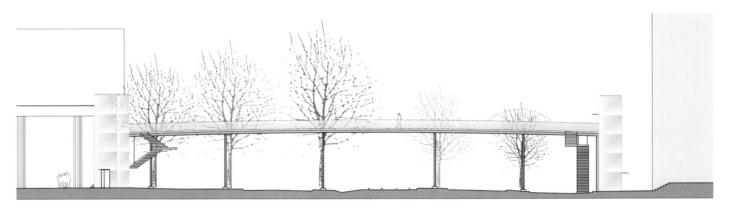

Haupteingang
Telekom Deutschland GmbH

accessible from the bridge deck.

The entire system is controlled by means of fibre optic cables from a media room in the adjacent underground parking facility. If the company network is accessed, the system can also be controlled from any given computer workstation. Owing to their particularly high luminance of more than 7,000cd/m², the video displays, which are also operated during the day, can be consecutively dimmed down to 30% of their maximum brightness.

These media bands display selected artistic contents. The extreme aspect ratio represents a particular challenge for video artists. The first media sequence of approximately 4 minutes consists of magenta squares and thus directly accesses an element from the Telekom logo. At irregular intervals shorter and longer chains of translucent squares move across the media ribbons. They can be interpreted as an allegory for information units overcoming by virtue of the bridge the arterial road which separates the company divisions. In the process, accumulations and tangencies occur, which lead to a summation of light intensities followed by discharged kinetic energy. The squares "break ranks", display arbitrary movements and react to encounters among themselves. The video functions as a metaphor for the fact that the exchange of information across data lines, just like the verbal knowledge transfer between Telekom employees using the bridge as a spatial connection, time and again releases surprising creative energies. This spontaneously emerging potential is often more valuable to an enterprise and the employees' self conception than the daily routine and repetitions of familiar work processes.

**White Paths and Stairs**

Good visibility on the bridge's circulation areas is provided by neutral white light from linear LED profiles. This lighting component is incorporated in all handrails and is thus—similar to the media screens—an integral part of the construction. A narrow beam lighting characteristic and the precise adjustment of the luminaires avoid road users and tram drivers being affected by glare when approaching or driving underneath the bridge. The strongly directional, powerful illumination of the very light coloured flooring on the footbridge and the flights of stairs creates a strong contrast with the relatively low lighting levels of the surroundings. Visually, the pathway is clearly detached from the urban environment; the footpath appears almost to be floating, which underscores the connecting

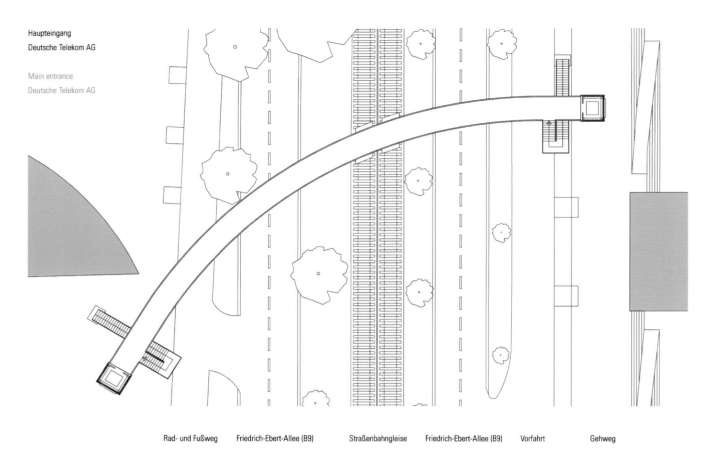

Haupteingang
Deutsche Telekom AG

Main entrance
Deutsche Telekom AG

Rad- und Fußweg    Friedrich-Ebert-Allee (B9)    Straßenbahngleise    Friedrich-Ebert-Allee (B9)    Vorfahrt    Gehweg

Lageplan der Infrastruktur im Bereich der Fußgängerbrücke

function of the construction between the building complexes.

### Backlighting of Elevator Shafts

The footbridge ends on both sides in 11-metre-high elevator towers. Owing to their glass cladding they emerge as airy stand-alone structures during the day, marking not only the beginning and end of the bridge but also the main entrance to the office buildings. In order to retain their functional dominance within the urban context during the hours of darkness, LED channels have been mounted behind the opal glass façades. The flexibly adjustable luminaires are located at the upper and lower edges of the towers and backlight their glass panels with neutral white grazing light. Through light reflections on the supporting structure inside the towers the spatial volume of the architectural constructions becomes readable. When in operation, the luminaires are dimmed to approximately 40% of their maximum light output in order to avoid disruptive glare from excessive vertical luminance levels.

During the dark morning and evening hours the tall glass cubes interact with passers-by. For this purpose LED profiles with very warm amber-coloured LEDs have been installed. Hidden beneath the shaft façades are double-pulse laser sensors which detect the presence and position of passing individuals and transfer these data to operating software. This activates the coloured light behind the vertical façade section nearest to the passing person. The distance between passer-by and tower determines the light intensity. Pedestrians comprehend intuitively that they are directing the lighting installation's dynamics. By coming to a halt or

changing direction they can influence the tower's lighting effect which is visible from far away. This experience of a physical "light-shadow" following the person gains yet another dimension when several pedestrians linger in the vicinity of the tower. The spatial relationships of the people to one another and of the individual to the group define the appearance of the tower.

**Technological Leap with Aesthetic Quality**

The Telekom Bridge in Bonn's Federal District is the first project in which the lighting designers of Licht Kunst Licht have made exclusive use of LEDs as the light source. The neutral white light-emitting diodes employed for the footbridge, stairs and elevator towers create an unusual, refreshing impression within the nocturnal urban environment, which generates high visual attention and aesthetic autonomy.

### 波恩德国电信步行桥——连接的纽带

德国电信股份公司是欧洲最大的电信公司,快速连接是其专业,经营了各种网络,通过电缆、卫星、无线电促成了人类和技术之间的沟通。然而有时,人类会更倾向于面对面地交谈各自的共同点。对于波恩电信总部的员工而言,这一层因连接繁忙街道两侧办公楼的步行桥的建设而变得更易实现。照明设计理念融合了互动元素,协调了整座桥梁建筑的华美风格,同时遵从了电信公司的口号:员工和公众得以通过这座桥感受"桥梁连接所带来的体验"。

72米长的桥对于在B9联邦高速公路两侧工作的4 000名电信员工而言可谓是福音,同时还为波恩的居民带来福祉。建桥前,行人需穿越弗里德里希·艾伯特大街(B9)这一段600米的距离,由此,许多人都试图抄近路,冒险穿越四车道和两条电车车道,安全是建造标高7米的步行桥的关键目的。舒适度是第二个目的,这促使桥两侧使用升降电梯的人流减少了。第三个目的在于建造企业形象的标识。大面积桥面仅仅由5根细高的钢管砼柱支撑,德国SBP工程公司(Schlaich Bergermann and Partner)设计的结构传达出现代主义和技术专精的理念。桥梁建设的照明理念还融合了电信公司的客户需求,如对企业形象标识颜色的依赖等。

### 品红色视频像素

品红带状照明无论是日间还是夜间都横跨了弗里德里希·艾伯特大街。桥梁纵向两侧的侧翼都安装了65米长的媒体显示屏,与整座桥梁相融合。这些LED视频显示屏面板的分辨率为32×6 500像素,色彩饱和度为16位。一系列显示屏模块无缝相连组合而成的媒体显示屏是为此项工程特别制定的。

在建筑师的设计过程中,最先考虑到媒体带状照明的整合。因此,LED模块和电子控制配件,主单元、线缆等都渐次整合进桥梁里。当电源切断时,媒体面板确实成为闲置的视频屏幕,却是桥梁结构不可分割的元素。所有的视频显示屏及系统配件都易于维护,从桥面即可操作。

整个系统由邻近地下停车库里的媒体房通过光纤电缆控制。一旦接入公司网络,任意一个电脑工作站都可以控制这个系统。视频显示屏的最大亮度超过7 000cd/平方米，白天也可以开启，可以连续调暗至最大亮度的30%。

带状媒体显示屏呈现了一些艺术元素。极端的宽高比对视频创作家而言是一大挑战。第一个媒体序列时长近4分钟，由许多品红方格组成，紧接其后便出现了电信的企业标志。较长或较短的半透明方格串不定期地出现在屏幕上，沿着媒体色带横向移动。它们寓意信息单位通过桥战胜了分离公司部门的主干道。在这个过程中，积累和相切相继出现，形成光亮度的叠加集合，进而释放动能。

"断序"的方格在显示屏上无序地运动，对迎面而来的其他方格做出反应。视频成为通过数据传输线进行信息交流的隐喻，如同电信员工通过桥这一空间连接线进行口述知识的交流，一次次释放出令人惊奇的创造能量。这一即时出现的潜力对于企业和员工的自我认知而言常常更具价值，远远超过日常熟悉的工作流程的重复。

### 白色桥路和阶梯

桥梁流通区域呈现良好的可见度，这得益于线性LED灯的中性白色光源。这种照明构件整合进了所有的扶手栏杆里，因此，与媒体显示屏类似，都成为桥梁结构不可分割的一部分。照明的狭窄光束和灯具的精确位置确保了道路使用者和重型机车司机的安全，使其靠近或驶进步行桥下方的时候不被眩目光影响。浅色桥面和阶梯具有强烈指示性的高度照明，与周围相对低伏的照明形成强烈的对比。视觉上，步行桥明显地独立于城市环境；同时，步行桥又似乎整个地漂浮起来，突显了桥梁连接办公楼的功能。

### 电梯井的背光照明

步行桥的两端各设一个11米高的电梯塔。玻璃外饰使电梯塔在日间仿佛一个空中独立的结构，勾勒出桥的起点和终点以及通往办公楼的主入口。为了保留城市背景里的夜间功能，设计师将LED显示屏装置在乳色玻璃外立面后上方。可调节的活动灯具装置在电梯塔的高处和低处，发出中性的白色光对玻璃面板进行背光照明。光折射到塔内部的支撑结构，建筑结构的空间体量便清晰可辨。当电源开启时，灯具的亮度渐渐调到最大亮度的约40%，进而避免了垂直照明的眩目光。

在天未亮之前和夜晚时分，高高的玻璃立方体与路人产生互动。LED灯设置为暖色系的琥珀色。电梯井外立面下方隐藏的双脉冲激光传感器能够探测到过往行人和车辆的出现以及方位，继而将这些数据发送到操作软件。由此，距离垂直外立面最近的过路人得以与外立面后方的彩色灯光产生互动。路人和电梯塔之间的距离决定了光的亮度。行人凭直觉便能感受到自身正在指挥着照明装置的动态效果。通过改变方向或静止停下，行人对电梯塔的灯效产生影响，而这一变化从远处

Detailschnitt Brückensteg
1. Handlauf, Edelstahlprofil
2. Lichtleiste mit LED 0,3W-Lampen
3. Glasbrüstung, VSG klar
4. Regenkanal
5. Revisionsöffnung
6. Brückensteg, Betonkern
7. Zu- und Abluftöffnungen
8. Videodisplays mit SMD LED-Lampen
9. Kragarm, Stahlprofile

就能望见。这一单个行人的物理"光影"体验在多个行人逗留电梯塔附近的时候将产生另一种维度。人与人之间以及个体与集体之间的空间关系决定了电梯塔的外貌。

### 技术跃进与美学

波恩联邦区域的德国电信步行桥是LKL公司的照明设计师首个仅采用LED灯源的项目。中性的白色发光二极管运用于步行桥的桥面、阶梯和升降电梯塔，为这个夜间的城市环境增添了别样的、焕然一新的景色，视觉魅力和美学特质都达到了极致。

# The Rion Antirion Bridge in Greece

## 神秘而壮观的梦中夜景
—— 希腊里奥·安托里恩大桥

**Credits**
Location: the Gulf of Corinth, Greece
Lighting Design: Roger Narboni, lighting designer, CONCEPTO agency (France), Sara Castagné, Frédérique Parent, project managers
Electrical Studies: Vinci Energies
Lighting Installer: Vinci Energies
Lighting Appliance Suppliers: Thorn and Philips Hellas (for the bridge illumination) – Schréder (for the road lighting)
Concession-holder: Gefyra
Bridge Constructor: Kinopraxia Gefyra
Architect: Berdj Mikaelian
Bridge Dimensions: 2,300m long, 27m wide, 4 towers rising 165m above seas level
Power Consumed by the Lighting: 400.8 kW, i.e. 7 W/m² (excluding tower surfaces)
Photography: documents Concepto

Night is falling on the Gulf of Corinth. The Rion Antirion Bridge joining the peninsula to the mainland is gradually metamorphosing. It now appears as nothing more than a thin golden thread, woven delicately through four towering bluish-coloured needles, a thread extending between two coastline fragments, a thread suspended between the sea and the heavens. The inky blackness of the water forms a dusky cradle, mirroring the towering icon that links the dark sea to the star-filled sky.

It is in this spirit that the illumination of this structure, whose construction began in 1999, was studied, to create a veritable nocturnal landscape matching the scale of this majestic site.

The illumination of the deck was intentionally designed to produce a super positioning of shadows bringing relief, texture and life to the immense length of the bridge (2.3 kilometres). The towers, illuminated with low-angle lighting from the bridge floor, gradually disappear into the dark sky. The piers and the sea are left in darkness to mirror the effects of the lighting.

**The Metal Bridge Floor – A Fine Golden Line of Light**

The two sides of the golden-yellow painted metal deck are illuminated laterally from above along their entire length by 560 THORN Contrast R1 floodlights equipped with Philips CDM-T 150W ceramic burner metal halide lamps (IRC 85, colour temperature 3,000K), a yellow glass filter and a barn door visor. All the floodlights on one given side of the deck are positioned in exactly the same manner and direction. Each floodlight is mounted on the end of the metal cantilever of the deck (every 8 metres) through its bracket and a slightly offset metal support, specially designed after on-site tests. They illuminate the deck laterally from a shallow top angle (in the direction of vehicle travel). The floodlights are set inwards from the end of the cantilever to make them as inconspicuous as possible. The exact height of each floodlight has been determined to create a horizontal shadow of the next cantilever on the side of the deck. The floodlight bracket was modified to minimize the surface area that can catch the wind. An indexing and blocking system has been provided to definitively set the orientation of the floodlights (in both the vertical plane and the plane of rotation of the floodlight on its bracket), including when replacing the lamps, which will be carried out from the front of the floodlight without removing the visor.

All the floodlights, supporting brackets and visors are painted in RAL 1018, the golden-yellow colour of the deck. Given that the bridge is situated in a maritime site with a highly-corrosive saline atmosphere, the appliances and accessories have received an anticorrosion treatment.

### The Bridge Towers: Bluish Needles

The bridge features four towers each supporting a 580-metre length of the bridge floor. Each tower comprises 4 pillars and rises 110 metres upwards from the bridge floor.

The four towers are lit at a shallow angle from the bridge floor level, with an intense bluish light that is visible from a great distance. Only three sides of each of the four tower pillars are illuminated with different brightness intensities in order to accentuate their parallelepiped shape. The fourth side, that is to say the interior side of the pillar facing the carriageway, is not lit. A total of 176 floodlights are used to illuminate the towers.

The outer side (facing the sea) of each pillar is lit up by 5 Philips Arena vision MVF 403 floodlights with ultra-intensive optics (2×3° aperture angle), equipped with Philips MHNSA 1,800W metal halide lamps (IRC 90, 5,600K) and featuring a special filter holder with a blue dichroic filter. The floodlights are positioned at the foot of each pillar and attached to the near-vertical concrete wall. They must be able to withstand winds of up to 200 kph and gusts of 56 m/s. The external lateral side (stay cable side) of each pillar is illuminated by 4 identical intensive floodlights, and the internal lateral side by 2 floodlights of the same type. Each floodlight has been precisely adjusted on site so that the light is projected evenly up the entire height of the illuminated side. The white stay cables - highly visible in daytime – become invisible at night, thus emphasizing the monumental scale of the 4 illuminated towers, which seem to be watching over the entrance of the gulf.

The road lighting, which also uses white light (metal halide lamps with ceramic burner), punctuates the structure and fits perfectly into the nocturnal stage-set.

The bridge illumination system consumes a total power of 400.8 kW (excluding the carriageway lighting). The lighting design takes into account the management and operating necessities to ensure the continuity of the nocturnal image. The chosen floodlights are catalogue products of tried and tested robustness and durability. The floodlight materials and protection indices are suitable for maritime

climatic conditions. All the lighting appliances are in readily-accessible positions on the structure to permit replacement of the lamps. The average lifetime of the 1,800W metal halide lamps is 4,000 hours (at 4,000 hours, 50% of the lamps have burned out), while for the 150W ceramic burner lamps it is 9,000 hours. The system lamps will therefore have to be changed every 2 years on average, with daily operation from nightfall until midnight (i.e. 2,000 hours of operation per year).

Inaugurated on August 8, 2004, the illumination of the Rion Antirion Bridge now permeates the nocturnal landscape of the Gulf of Corinth. The thin golden thread that joins the Peloponnesus peninsula to the mainland seems to be holding together the coastal fragments which, situated on a tectonic fault, are moving inexorably apart by 8mm per year.

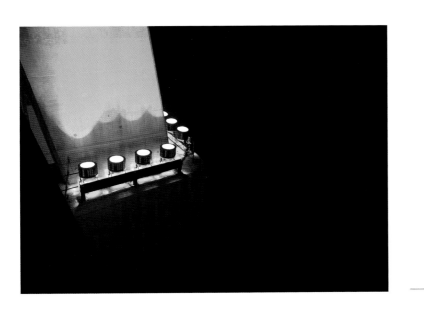

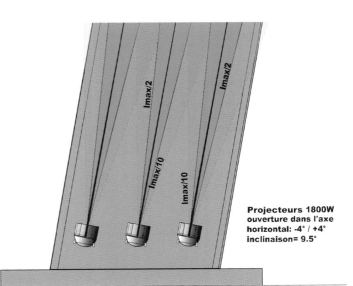

**Projecteurs 1800W
ouverture dans l'axe
horizontal: -4° / +4°
inclinaison= 9.5°**

夜幕降临在科林斯湾。连接着半岛和内陆的里奥-安托里恩大桥的影像悄然变化，仿佛只是一根金色的细线，雅致地将四根高耸的蓝针织在一起，这条线的两端分别延伸至滨海陆地，悬挂在天空与海洋之间。黑漆漆的水体形成灰蒙蒙的摇篮，高耸的物象倒映在黑色的海面，将海洋与繁星点点的天空相连接。

这一景致浓缩了大桥照明的精髓，其施工开始于1999年，力图营造与场地庄严的尺度相呼应，夜景富于变幻的景观。桥面的照明设计在于创造超大尺度的影子，将放松、质地和生命注入这座跨度超长的大桥(2.3千米)。桥塔采用低角度照明，从桥面逐渐向上直至消失进黑色的天空里。桥墩和海洋在黑色里反射着照明的效果。

#### 金属桥面——金色细线照明

金黄色喷漆的金属桥面的两侧，整座桥的侧边靠上处装置了英国索恩牌560系列的泛光灯，同时配以飞利浦CDM-T型号150瓦陶瓷燃烧金卤灯(颜色还原IRC 85，色温3 000开尔文)、黄色玻璃片和挡光板。桥面侧边装置的所有泛光灯的位置和方向都一模一样。

每个泛光灯装置在支架和偏置式桥梁支座之间，处于金属悬臂的桥面一端(悬臂的间距均为8米)，这是经过现场测试后而特别设计的。泛光灯内置在悬臂末端，尽量使其显得不起眼。每个泛光灯的高度都十分精准，将桥面一侧邻近的悬臂的影子营造为横向型。

泛光灯的支架表面尽量小化，使其不迎风。标引和闭塞系统用于固定泛光灯的方位(竖向平面和泛光灯在支架上的旋转平面)，如：无需移动挡光板就能从前端置换灯泡。

桥面采用色卡RAL1018的金黄色，所有的泛光灯、支架和挡光板也都采用此色系。即使桥梁处于高度腐蚀的盐性海水区域，这些装置和灯饰也早已做过防蚀处理。

#### 桥塔——蓝针

桥梁的特色体现在四座桥塔，每座桥塔都支撑着580米长的桥面。每座桥塔由4个塔柱组成，从桥面向上延伸110米。

四座桥塔被强烈的蓝光点亮，即使在很远的地方仍能望见，光与桥面所成角度十分微小。每根塔柱的三个侧面采用不同的亮度来突显其平行六面体的外形。剩下的那个侧面，即塔柱面向行车道的内侧，则不采用照明。桥塔所用的泛光灯总数达176个。

每根塔柱的外侧(面向大海)采用5个飞利浦MVF 403高功率泛光灯，带有超精密透镜(孔径角2×3°)，配以飞利浦MHNSA型号1800瓦的金卤灯(IRC 90，色温5 600开尔文)和用于蓝色二向色滤光镜的特制滤光片夹。泛光灯装置在每根塔柱的根部，附着于近垂直的混凝土墙体。这些泛光灯必须能够抵抗每小时200千米的风和每秒56米的狂风。每根塔柱的外部横向一侧(大桥斜拉索一侧)装置了4个相同强度的泛光灯，对侧则采用2

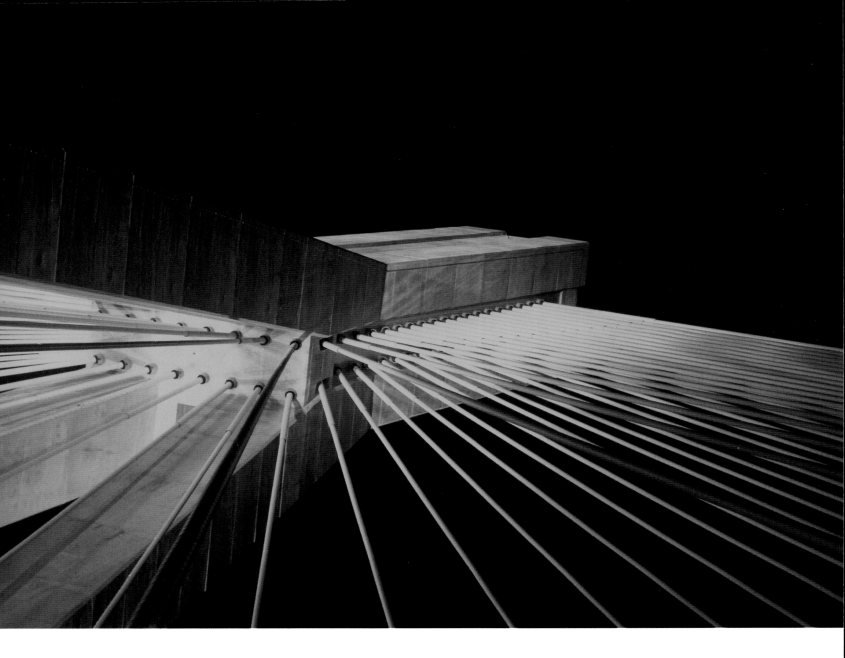

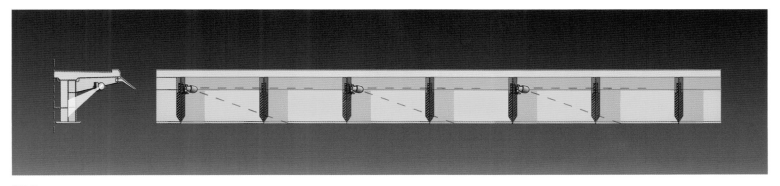

个相同型号的泛光灯。每个泛光灯的位置都十分精确，保证灯光能够向上照亮整个侧面。白色的斜拉索，白天清晰可见，夜晚却遁形无踪，以此强化4座被照亮的巨大桥塔，仿佛它们在守望着科林斯湾的入口。

桥面照明采用白色灯光（陶瓷燃烧金卤灯），由此点缀着整个桥梁架构，完美地融入整个夜间场景。

大桥照明系统的总耗电量为400.8千瓦（除了行车道照明）。照明设计整合了必要的经营和管理元素，确保了夜景的连续性。选用的泛光灯是稳定性和耐久性经过测试和试用的目录中的产品。泛光灯材质及其保护指数都适用于海洋性气候条件。所有的照明设备都易于更换灯泡，大桥结构上的装置位置方便可达。1 800瓦金卤灯的平均寿命为4 000小时（如果达到4 000小时，那么50%的灯泡都会烧坏），150瓦陶瓷燃烧金卤灯则为9 000小时。因此，此照明系统的灯泡必须两年一换，日常运行时间从傍晚至午夜（即每年的运行时间为2 000小时）。

里奥-安托里恩大桥建于2004年8月8日，现今，大桥的照明弥漫进整个科林斯湾的夜景里。金色细线连接了伯罗奔尼撒半岛和希腊内陆，似乎将将地质断层上的滨海地块全部整合在一起，这些地块正以每年8毫米的位移相分离。

# Current³

## 会"呼吸"的互动性光照设计
## ——Current³

**Credits**
Light Artist and Concept: Virginia Folkestad, sculptor
Lighting Design Technical Consultancy:186 Lighting Design Group, Inc.
LoDo Lights Project Director: Diane Huntress
Electrical Installation: Rumbold Electric
Special Thanks: Sponsors - Rudi Cerri, Denver Arts and Venues and Lower Downtown Neighborhood Association
Photography: Diane Huntress Photography

Current³ by Colorado artist, Virginia Folkestad, is the first Denver light media installation in the LoDo Lights project inspired by Turin.

Beginning in 2007, Huntress, a photographer, began the approval process to create a project and fund contemporary light art media on a 1908 steel railroad truss repurposed into a pedestrian bridge in the Denver Historic District, LoDo. The public-private collaboration was initially planned to be in place about 60 days. Project Gallerist and Art consultant was Carol Keller. The Lower Downtown Neighborhood Association a non-profit, became the fiscal conduit. The Denver office of Arts and Venues brought in final funding and the public jury process to select the artist. The public dedication was March 6, 2009 and honored the 150th birthday of Denver, Colorado USA.

Public sentiment and more negotiations have kept the popular installation ongoing and interacting with passers-by. The bridge connects downtown Denver to the Pepsi Center sports arena and popular bike trail.

Current³ by sculptor and installation artist, Virginia Folkestad, is a site-specific, interactive light

installation that encourages play and collaboration. Installations modify the way a particular space is experienced and this installation is meant to give users of the Wynkoop Street Railroad Bridge an immersive experience; a change from the usual, familiar occurrence of traveling across the bridge. Passersby become aware of their place in time and space.

As with many examples of Installation Art, this work invites the visitor to take control and become a participant in the completion of the work by turning the bridge into a kinetic sculpture. The light that

appears in response to movement, invites one to explore further by continuing to cross the bridge. Concurrently, projections of drawings onto the deck of the bridge give a nod to its rail history and it's significance in the development of Denver.

When at rest and no pedestrians are on the bridge, Current [3] "breathes". The Bridge exhales as the RGB LED color projectors dim and shift from a blue to a lavender, then inhales again as the lights increase in intensity and turn back to a saturated blue. The timing of the breathing was set on site to match the rhythm of someone walking and breathing. People can sync with the breathing of the installation as they approach and step onto the bridge. As people cross the bridge, electronic trip sensors trigger events. Light patterns with artist designed gobos are projected for short bursts, while the color projectors change colors and move from the breathing mode to the static color mode for a pre-determined time period. With a single person crossing the bridge, the effects follow them before returning to the breathing mode. The last event is triggered by a trip sensor near the edge of the bridge, encouraging people to turn around and look back at Current³ as they exit the final event. The installation interacts with groups of people much differently than it does with a single person. As people walk both directions or in groups, the event triggers are reset at varying times such that Current³ gives each visitor their own unique experience.

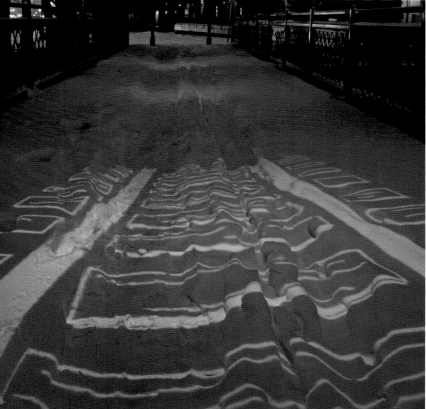

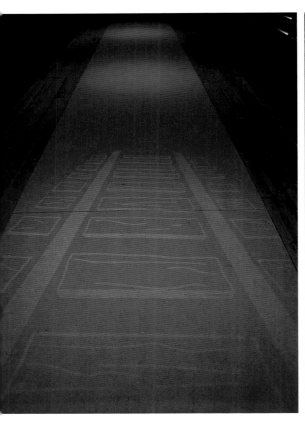

Current³是科罗拉多当地艺术家费吉尼亚·福克斯塔德的手笔,它是丹佛第一座媒体照明设施,是下城中心历史遗产建筑群照明工程的一部分,被都灵人广为传颂。

2007年,由摄影师享奎斯牵头,翻新过程开始了,她在1908年建造的铁路桁架上安装现代的照明设施,将其改造成一座人行天桥,桥梁位于丹佛历史悠久的下城中心区。设计时花费了约60天时间对公共&私有建筑进行规划。项目美工及艺术顾问为卡罗尔·凯勒。下城社区委员会这样的非盈利性机构,成了项目资金中间方。丹佛场地艺术办公室则是资金提供方,并通过大众评选出艺术家。2009年3月6日进行了公共投票,以纪念美国科罗拉多州丹佛的150岁"大寿"。

公众的情感和热烈的讨论推动着项目的进行,并使旁观者参与其中。桥梁连接丹佛中心城区、百事可乐中心及公共自行车车道。

Current³出自Virginia Folkestad事务所的雕塑家和建筑艺术家之手,他致力于营造位点专一的互动性光照设计,从而增强娱乐性和协同性。装置改变了一个特定空间被体验的方式,也注定为那些乘坐温库普街铁路桥的人们带来别样的体验;这里原先是一条穿过桥梁的道路,十分普通、平常。现在行人每时每刻都关注那个地方。

如同装置艺术的许多例子一样,桥梁可以变化成雕塑,以便由寻常百姓来掌控,并参与到项目建筑的队伍中来。而灯光作为对此的回应,邀请人们穿过桥梁,继续究微探秘。同时,位于桥梁甲板上方的投影图则确证了那里的零售史,这对丹佛的城市发展意义重大。

休息时间,桥上无人,Current³开始"自然呼吸"。当RGB LED彩色投影机变暗并从蓝色转为淡紫色,则为呼气状态;当灯光强度增加,且变回蓝色时,则为吸气状态。呼吸的时点被设定成与人们行步和呼吸的节奏匹配。当人们接近并踏上桥梁时,可以尝试和灯光同呼吸。当人们过桥后,电子传感器会开始活动。图案投射灯投射出短脉冲的灯光,而彩色探照灯则变化着颜色,并随着预设的时间,在呼吸模式和静止模式间来回转换。一旦有人穿过桥梁,灯效在变为呼吸模式前,会随行人的呼吸变化而变化。而传感器启发的最后一个活动在桥边附近,让人们在离开之前再度回望Current³。建筑影响了不同人群,而非单一个人。当人从桥的两头,独自或成群结队地进入桥面,活动触发器都会进行不同设定,使每位游客都能从Current³带走不同的体验。

# Swivel Bridge, Bremerhaven

## 间接照明的典范
——不来梅哈芬平旋桥

The new pedestrian bridge links the city center with the Havenwelten. Designed with a swiveling center part of 42m, it allows ships the passage to the museum harbor basin. The entire structure is illuminated with an indirect, glare-free lighting system. Integrated in the floor, frosted glass fields illuminate the inner glass shell, which is printed with a pattern of matt white dots – the light reflects smoothly into the space.

这座新步行桥起到连接市中心和不来梅哈芬的作用。平旋桥的旋转部分长42米，船只可以从这里驶进博物馆港湾。整个结构采用间接照明系统，完全没有眩目光，桥面的磨砂玻璃铺装部分内置灯具，照亮了上方的玻璃内壳，印有哑光白点的图案折射出柔和的光，笼罩了整个空间。

**Credits**
Location: Bremerhaven, Germany
Client: BEAN sellschaft, Alter / Neuer hafen mbH & Co KG
Length: 100 m
Lighting Designers: pfarré lighting design
Architects: NPS Tchoban Voss
Engineers: WTM Engineers GmbH
Photography: Markus Tollhopf

# Epping to Chatswood Tail Link

精炼与提纯
——艾平－查茨伍德换乘平台

**Credits**
Location: Sydney, New South Wales, Australia
Public Area: approximately 6,000m² per Station
Architect: Hassell
Services Engineer: WSP Lincolne Scott
Photography: Brent Winston Photography
Lighting Design: PointOfView

The Epping to Chatswood Tail Link is a rail extension comprising four stations that link rapidly growing suburbs north west of Sydney with the commercial hub of Chatswood, and through to the Sydney central business district.

The lighting brief was to provide innovation to create safe, attractive, efficient and durable subterranean spaces that encourage public use other factors which informed the lighting design were the need to counteract natural human phobias about being underground and to create a strong visual signature above ground.

Despite the scale of the spaces and the need for perceptibly bright environments, an average lighting load for public spaces of just 9.8W/m² was achieved through innovative design, exacting testing, development of bespoke fixtures, and the judicious application of light.

### Context
The diverse project objectives are satisfied through cultivating a design that balances lit and unlit surfaces, and creates a perception of brightness from minimal use of light. Rigorous refinement condensed the architectural lighting to an essential scheme that satisfies compound technical, practical, creative and emotional challenges. This distillation created a purity of design in which nothing is superfluous; each fixture has a specific task, and functions without waste resulting in a superior spatial experience.

### Ground Level
By day natural light fills the street entry cabin which is approximately 30 metres deep. At night the station canopy presents as a crisp luminous beacon; up-lighting to the canopy creates a luminous envelope which provides a dramatic signature for the station. This "glow" serves to also illuminate the station apron,

virtually eliminating the need for supplementary area lights, and additionally to alleviate the oppressive sensation of entering a deep hole at night. By reflecting light from the fritted glass the effect of daylight is mimicked after dark; the ceiling is bright during both day & night.

### Ticketing Cabin

In the huge ticketing cabin, light reflected asymmetrically to the space provides the primary illumination; the biased application of light assimilates with the structure's asymmetric form. This effect is achieved through the development of a bespoke bracket fixture. This custom fitting is effectively an IP65 housing with a range of "cassette" optics; each fixture module is 3m long, to suit the station building module. Down lights supplement lighting to the centre of the space sufficient to provide average ambient illuminance of 100lux in the thoroughfare (though the

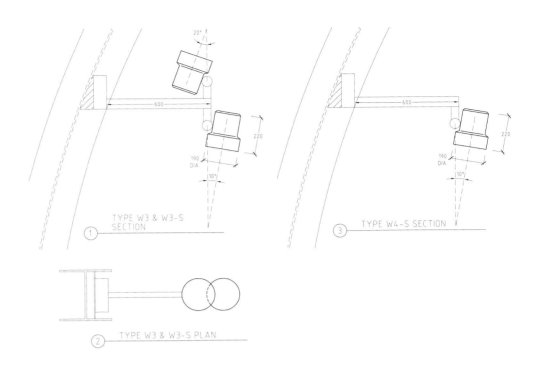

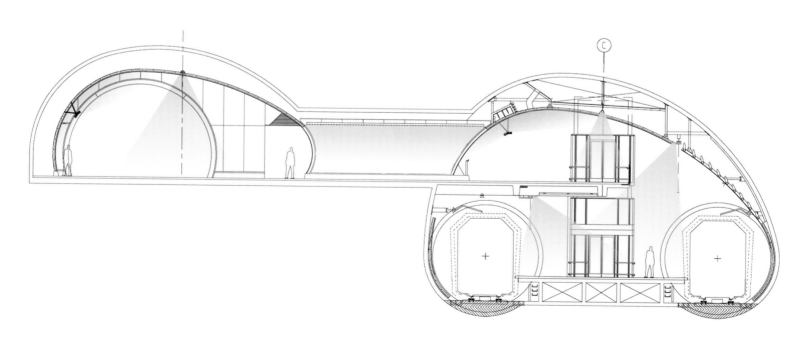

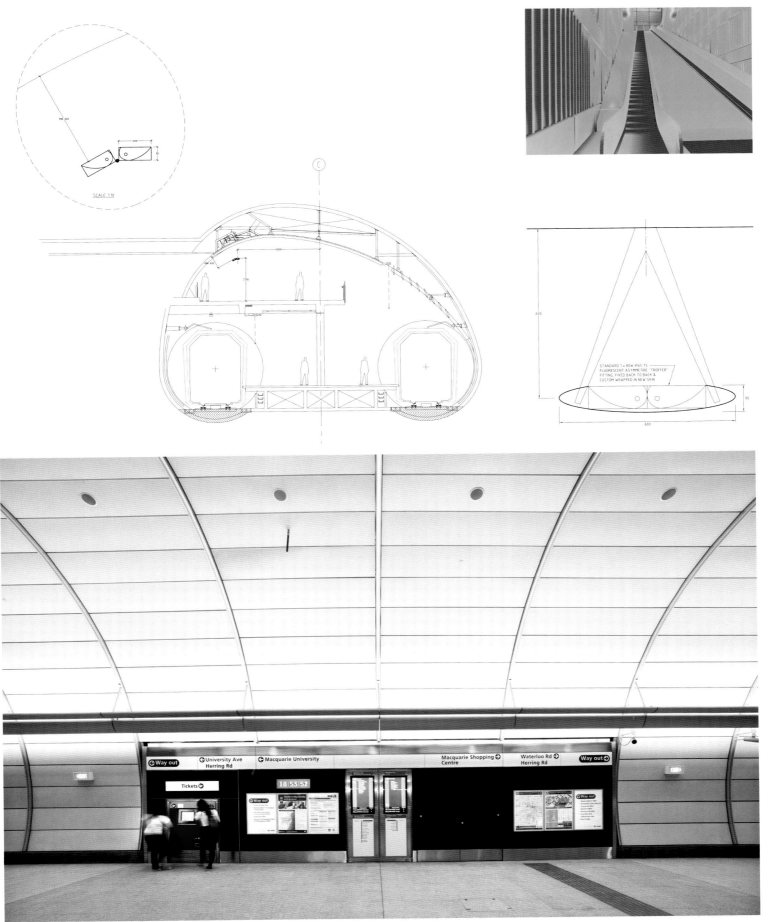

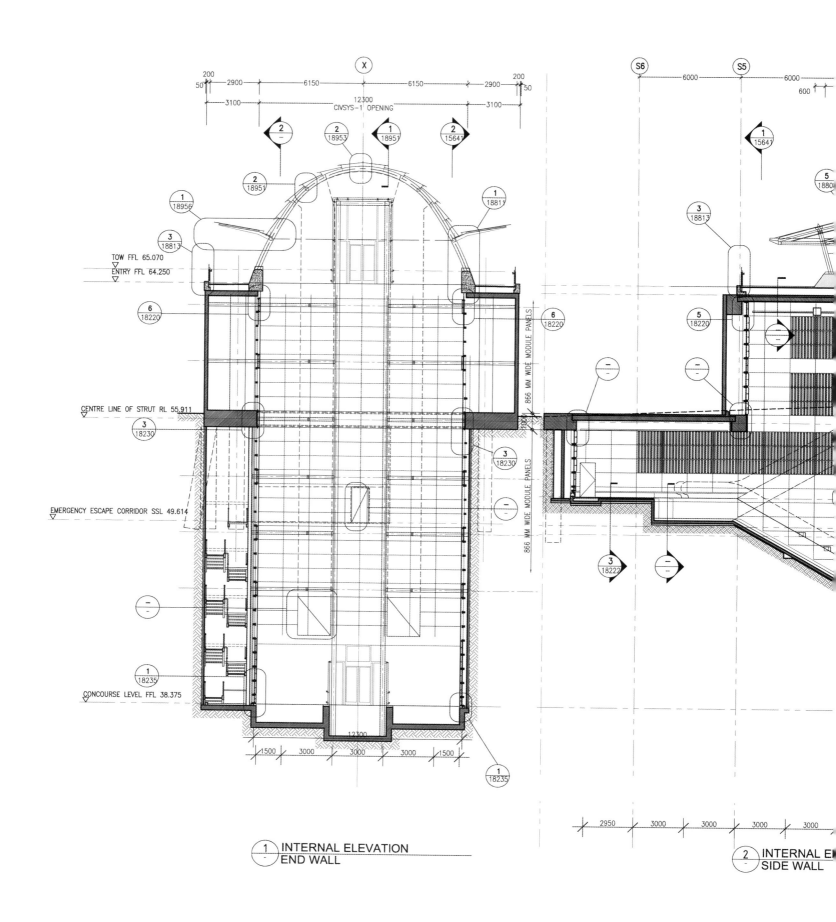

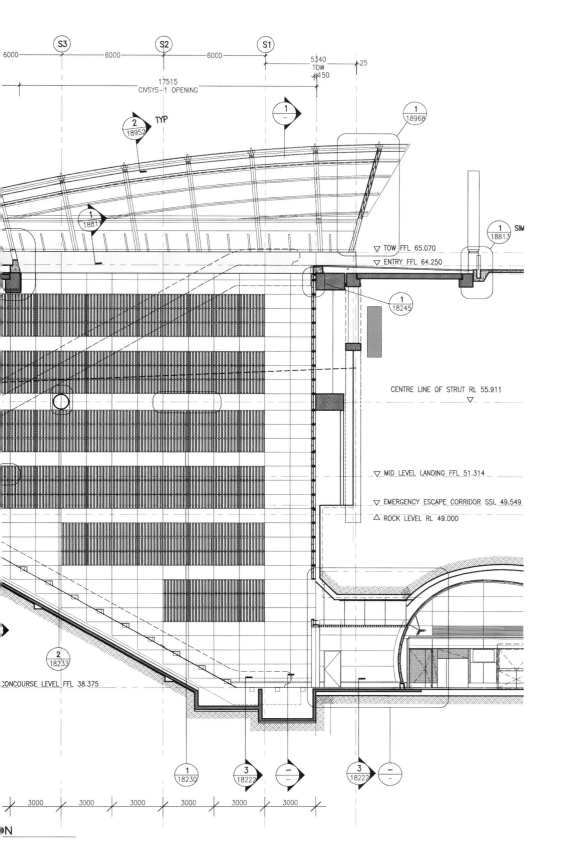

impression is of a significantly brighter space). As the journey develops, commuter orientation and flow is assisted by the use of different colour temperature light; warmer light and enhanced levels (300lux) enhances the turnstiles which subliminally provides hierarchy.

### Platform Cabin

In the platform cabin a minimum 200lux was required at the platform edge for obvious safety reasons. The challenge in this large and open volume was to create the impression of a uniformly bright and airy space without the lighting equipment being obtrusive. Effectively, just 2 different fittings are used for primary illumination; the custom bracket fixture which indirectly illuminates the volume at high level, and a linear modular system that directly lights the space from above each platform.

The impression is of a naturally luminous space. Due to the harmonisation of lighting with the physical form and attention to simplicity, lighting products have only minimal visual presence whilst providing a welcoming, safe experience that counteracts any sense of entombment.

艾平－查茨伍德尾部转换平台是一个由4个站台组成的铁路配套设施，连接快速发展的悉尼西北郊外和查茨伍德商业中心，并且通向悉尼中心商业区。

照明设计的理念在于使用创新手法来营造安全、引人注目、高效和持久的地下空间，使公众在利用照明设计消除在地下的自然恐惧之余获得其他体验，以及在地上创造一个醒目的地标。

尽管空间较大且有着对于明亮环境显而易见的需求，通过创新设计、严格的测试、定制安装的发展和对光线的明智选用还是实现了每平方米公共空间平均9.8瓦的光辐射度。

### 环境

通过平衡光照和未被光照的表面能实现不同的项目目标，用最少的光照来营造明亮的视觉感受。严格的细化将建筑照明精简为一个满足复合技术、实用、创造性和情绪挑战的精华方案。这种蒸馏法过滤了设计中所有多余的部分，每个设备

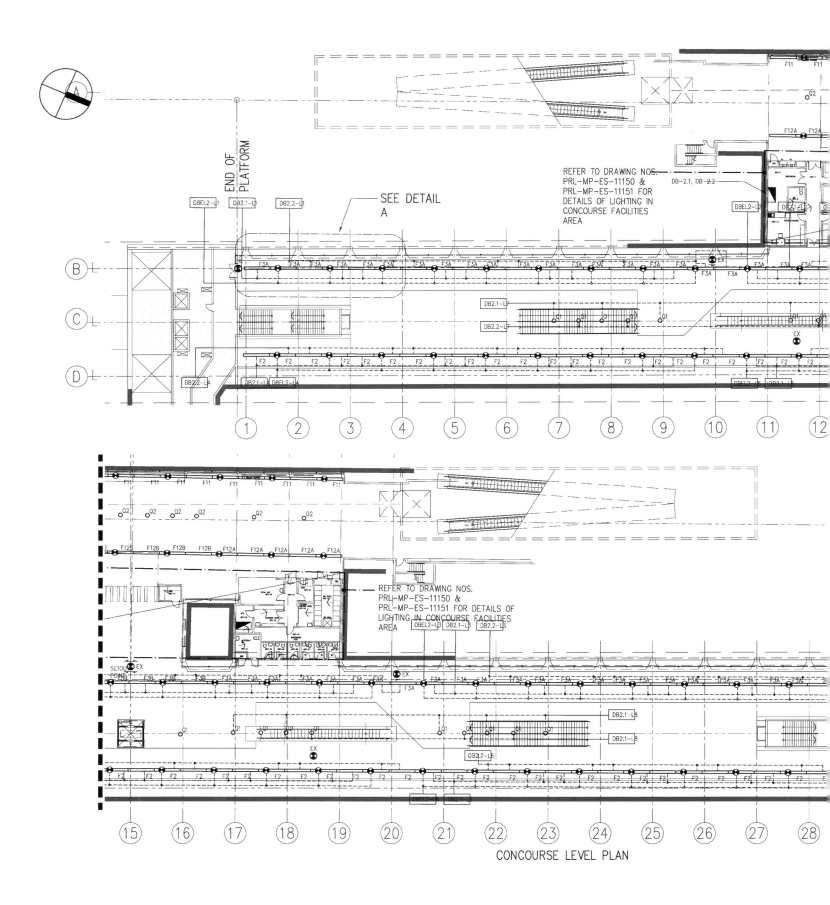

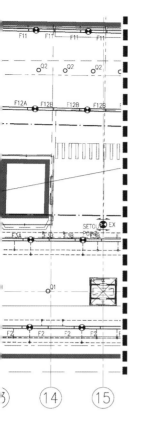

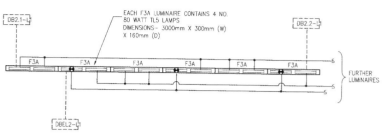

DETAIL A – TYPICAL LUMINAIRE WIRING ARRANGEMENT
CONCOURSE LEVEL

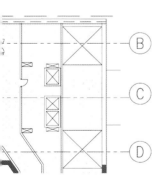

都有专门用途，营造出出众的空间感。

### 地下层

白天自然光充满了这个约30米深的街道入口信号房。夜晚站台天篷变成了一个弯曲的发光塔，从下往上照亮天篷的灯光创造出一个发光的外壳，为车站提供了引人注目的识别标志。灯光也照亮了站台，实际上消除了对增补区域光源的需要，此外也减轻了在夜晚进入一个深洞的压迫感。利用烧结玻璃的反射光在入夜之后活灵活现了日光效应，天花板在白天和夜晚都是明亮的。

### 售票舱

在巨大的售票舱里，不对称的光照为空间提供了主要照明，这种偏向的应用对应了建筑不规则的形式。这效果来自于一个定制的支架夹具。这种定制的夹具实际上是一个带有一套卡带式灯具的IP65等级的框架。

中间的向下补充灯光给通道提供了平均充足的100勒照度(看上去是一个引人瞩目的更亮的空间)。

继续向前使用了不同色温的灯指示乘客的方向与人流，更暖和的光线和加大的照度(300勒)强化了十字转门下意识的分级作用。

### 站台舱

出于安全需要，站台边缘至少需要200勒的照度。在这个开放的大空间里创造出灯火通明和通风的一致空间感却不突出照明设备是设计的难点。实际上主要照明仅使用了2种不同的设备，上层是间接照亮空间的定制支架固定装置，每个站台上方是直接照亮空间的线型模块系统。

最终呈现的是一个自然光照的空间。由于照明在外形上的和谐以及对简朴的注重，照明产品的视觉存在是最小化的，提供了一种友好，安全的体验。

# Los Angeles World Airports

黑夜只是背景
——洛杉矶国际机场

**Credits**
Location: Los Angeles, USA
Lighting Design: Selbert Perkins Design
Other LAX Beautification Enhancement Project
Team Members: Ted Tokio Tanaka Architects, IMA Design Group, Lighting Design Alliance, Isenberg & Associates, Inc, Ove Arup & Partners

Selbert Perkins Design (SPD) was chosen by the City of Los Angeles Department of Airports to create a new identity for the Los Angeles International Airport (LAX) and comparable identities for Palmdale (PMD), Ontario (ONT) and Van Nuys (VNY) — the area's other three aviation centers. The new identity functions as a "system" of logos that work for all four airports and each one individually thereby uniting the airports and promoting confidence.

Following the graphic identity work, SPD was also selected as the environmental design consultant for the LAX Beautification Enhancement project. The mission of the project was to develop and implement a plan to greatly improve the identity, function and circulation of the airport through the use of environmental communications, sculpture, architecture, landscape design, lighting and art. This project extends from Century Boulevard in and around the LAX site and up to the doors entering the airport.

The most dramatic result was SPD's dynamic landmark gateway into LAX. The gateway includes 32-foot high LAX letter forms, a ring of fifteen 120-foot pylons forming a bold gateway into LAX, as well as equally-spaced columns that steadily increase in height along a two-mile roadway (Century Boulevard) median leading up the gateway. The nighttime personality of the gateway dramatically changes from its daytime

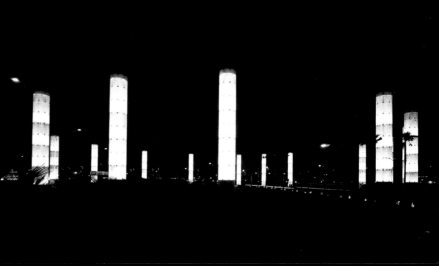

appearance through the use of computerized lighting effects and airline passengers can begin to view them from 3,000 feet above. This gateway ring design was symbolic of the "City of Angels" with the multiple colors representative of the multi-cultural city, being one of the most diverse in the nation.

凭着出色的设计能力，瑟尔勃特设计公司(SPD) 被洛杉矶航空局选择为洛杉矶国际机场(LAX)设计新的企业标志，同时还为该地区的其他三个航空中心Palmdale (PMD), Ontario(ONT) 和Van Nuys(VNY)设计特征性标志。新标识可以作为四个机场的一套logo "系统" 使用，因此将机场统一了起来并促进了信任的建立。之后SPD还成为了LAX美化项目的环境设计顾问，协同LAX的相关人员共同致力于整个机场的环境改善。美化项目的目的在于为旅客创造更好的环境，在旅客的来来往往中为他们提供更多的便利。为了实现这个目的，设计人员进行了多方面的工作，例如，对机场的整个环境网进行了改善，适当地布置了雕塑，对机场内的建筑和景观进行了合理的规划和设计，此外，还对灯光和其他艺术创造进行了巧妙设置。设计范围非常广泛，从世纪大道开始，一直绵延到机场的入口，遍及LAX机场的里里外外。

在设计师的精心设计和创作下，整个机场呈现出夺目的光彩，尤为突出的是入口处的动态陆标。高达9.8米的三个字母LAX成为机场显眼的标志之一，还有15根高为36.6米的路标塔，它们组成一个环形，高大而壮观。沿着3.2千米长的世纪大道边设置了许多圆柱，它们的高度规则增加，犹如路标般将旅客引入机场。

当夜晚来临的时候，入口处完全呈现出另外一种风景，凸现在人们眼前的不再仅仅是白天那凝固的静态之美，随之增加的是灯光交织中的动态和妩媚。高高的路标塔的灯光全部都亮了起来，即使旅客在900多米的高处也可以看到这耀眼的光塔。在电脑的控制和调节下，这些路标塔变换出不同的颜色，一方面为整个机场增添了眩目的色彩，另一方面又象征了洛杉矶的多元文化。在色彩的交替下机场的入口显现出一片灿烂和魅力，让"天使之城"这一意蕴在迷人的景致下展现得淋漓尽致。

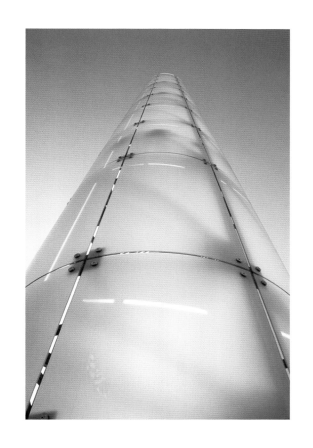

# Underground

## 多种光源在市政建设中的运用
——科恩路地道

**Credits**

Client: Downtown OKC, Inc.
Scale: 3,901 m²
Architect: Elliott + Associates Architects
Project Team: Rand Elliott, FAIA
 Michael Hoffner, AIA
 Joseph Williams, Assoc. AIA
Historian: Pendleton Woods
Cost: $1,277,333
Photography: Robert Shimer, Hedrich Blessing
Scott McDonald, Hedrich Blessing

The Conncourse, named for Jack Conn, is a system of tunnels and bridges which connects 23 buildings in the core of downtown Oklahoma City. The first tunnel was constructed in 1931, crossing under Broadway Avenue and connecting the historic Skirvin Hotel to Skirvin Tower (now known as 101 Park Avenue). The majority of the Conncourse was built between 1972 and 1984.

Virtually untouched since the 1970s, the space had fallen into disrepair with duct-taped carpet and brown-on-brown color scheme. Through the leadership of Downtown OKC, Inc., the City of Oklahoma City and the Conncourse Association, a plan was developed whereby the "attached" building owners would assess themselves in order to make the improvements. The portions of the Conncourse that traverse buildings are owned and maintained by attached individual building owners. The portions of the Conncourse that are between buildings are maintained and managed by Downtown OKC, Inc.

The purpose of this project is to improve the quality and public perception of the Conncourse. Changing the name to the Underground is one step in changing the perception. The combination of music, colored light, understandable way-finding and historic themed galleries make moving from point A to B rhythmic, surprising and educational. The re-invention of the Underground has transformed the space from a dark, dated maze to a "walkable" entertainment. A new program supporting art installations, in addition to the historic Downtown galleries, focuses on art in vacant storefronts to create interest for new retail. There are mile markers installed in order to allow downtown walkers to calibrate distance.

Signage and a new logo, also designed by the Architects, and way-finding elements play a crucial role in the success of the Underground. Outdoor signage at Underground entrances is consistent and easily seen from the street. New brightly colored "Portals" and "Panels" mark tunnel intersections and incorporate signage to help users locate their destinations. The Underground concept uses color as a navigational tool for users. The security guards give directions by suggesting that to get from point A to point B, you simply follow the green light to the red light to the yellow light to reach your destination.

Underground segments adopt "gallery" themes to make locations memorable, helping visitors navigate the space and providing visual interest along the way. Photographs and art relating to the nine themes

are hung on gallery walls, providing points of interest and opportunities to learn about Downtown Oklahoma City's history. The historical photo galleries are "Banking in Downtown Oklahoma City", "City and County Government", "Downtown Commerce", "Energy and Public Utilities in Downtown Oklahoma City", Our Federal Government Downtown", "The Murrah Building", "Construction of the Federal Courthouse", and "Overview of Downtown History". The "Light Gallery", one of 11 different colored lighting installations, provides users with the unique experience of walking through a permanent art installation. The Light Gallery uses blue and yellow light to create white light.

# 04 Cityscape & Public Art
城市景观+公共艺术

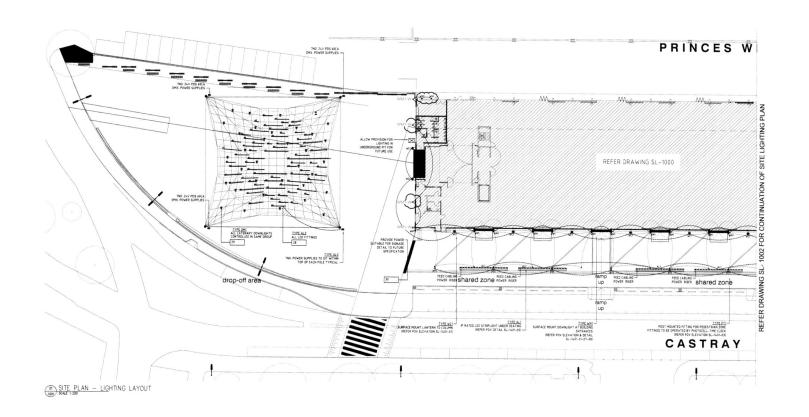

| | | |
|---|---|---|
| ■ | ——Clichy Batignolles Park in Paris 17e<br>巴黎17区克利希巴蒂诺尔公园 | the lighting was purposefully designed on high environmental quality. Other parameters included care not to disturb the animal and plant life and a determination not to increase light pollution in Paris.<br>照明设计定位基于可持续发展和高品质的环境质量。其他纳入考虑的因素：不妨碍动植物的生命，不增加巴黎的光污染。 |
| ■ | ——Robert Kennedy Inspiration Park<br>肯尼迪纪念公园 | The team chose to not illuminate the trees themselves in order to minimize light pollution and to emphasize the tree profiles.<br>设计团队选择不照亮树木，是为了减少光污染，和强调树木的外观。 |
| ■ | ——An Eventful Path<br>悉尼奥林匹克公园历史大道 | The linear path is an abstract time line, which celebrates the past and looks to the future.<br>这条直线形的道路象征的是一个纪念过去，展望未来的抽象的时间列表。 |
| ■ | ——Slater Mill Falls<br>斯莱特米尔瀑布 | It was important that the mills be seen and the fixtures not.<br>看得见的厂区和看不见的照明设备是两处重点。 |
| ■ | ——Shed 1, Princess Wharf<br>公主码头1号棚 | The essence of the lighting scheme is to accentuate the building as a backdrop for the various functional demands, and also as a feature element to the local environment at night.<br>照明设计的本质在于强调建筑作为满足各种功能要求的交流屏，也是夜晚当地环境的景观元素。 |
| ■ | ——The Port Pavilion<br>圣地亚哥百老汇港口大厅 | Tidal Radiance, was designed in concert with an environment of projections to evoke tides and sea life.<br>"潮汐"外立面雕塑，通过投影于所处环境，唤醒人们对潮汐和海洋生活的回忆。 |

# Clichy Batignolles Park in Paris 17e

## 超低能耗的特殊照明
——巴黎17区克利西巴蒂诺尔公园

**Credits**
Location: Paris, France
Client: Paris City Council, Parks Department, Jacqueline Osty agency, landscape Architect: François Grether, architect.
Lighting Design: Roger Narboni, CONCEPTO agency, Frédérique Parent, project manager
Electrical Studies: O.G.I.
Installer: CEGEX
Cost of Lighting Work: 340,000 Euros excluding VAT (2005 value)
Total Power Consumption: 9 kW
Materials Used: Aubrilam Volta 1 independent poles with photovoltaic cells and 10 x 1W blue LEDs - Comatelec Phylos LED light columns with 20 x 1W white LEDs, 3000K - Sermeto Carène mast with 2 Thorn Contrast projectors, IMC lamp 70W, 3,000K – devices embedded in the ground and into water: LEC with 1 x 1W LED, 8,000K.

The lighting in Clichy Batignolles Park has been designed with the following aims:
- To create various, attractive, nocturnal atmospheres
- To increase the safety and security of the main paths
- To provide a visual limit to the park, which lies adjacent to large railway yards
- To improve the park's night-time image since it is visible in urban landscapes and from neighbouring blocks of flats

Because the site was initially looked at as the location of the Olympic village when Paris was still a candidate city, the lighting was purposefully designed on the basis of sustainable development and high environmental quality. Other parameters included care not to disturb the animal and plant life and a determination not to increase light pollution in Paris.

### Providing a Visual Limit to the Park
Given the very unusual position of the park, bordered on the west side by a large rail yard, it was important, indeed vital, to provide a visual boundary, separating the park from the inaccessible areas beyond it. Along the western side, the ends of the network of secondary paths are indicated by small white lights equipped with 20×1W white, low-consumption LEDs mounted on top of cylindrical metal masts 5 metres high visible from some distance away.

### Lighting the Inner City Circle Rail Line
In accordance with the Lighting Master Plan for the Paris "Crown", and for the first time, the small "circle" rail line that runs in a wide curve past the basins at the northern end of the park has been underlined by lighting. It was highlighted by blue lights mounted on wooden posts six metres high and powered by independent photovoltaic cells. The posts have been set up along the bank at a regulation distance from the railway tracks but are increasingly widely-spaced. They culminate in the centre of the basin to accentuate the curve effect.

Each light is equipped with a photovoltaic panel and a battery built into the post, making it completely independent as regards power supply. This means that it does not require any civil engineering work (trenches, sheathes and cabling) other than its concrete foundation. Because of this, the lights are very easy to move, if required in later modifications and future layouts scheduled for this area.

### Night-time Displays
The themes of the body, the seasons and plants developed by the landscaper in the centre of the park each include a night-time image. The observation platforms overlooking the park are lit with button lights consisting of 1×1W LED. These carefully-positioned night-time displays add depth

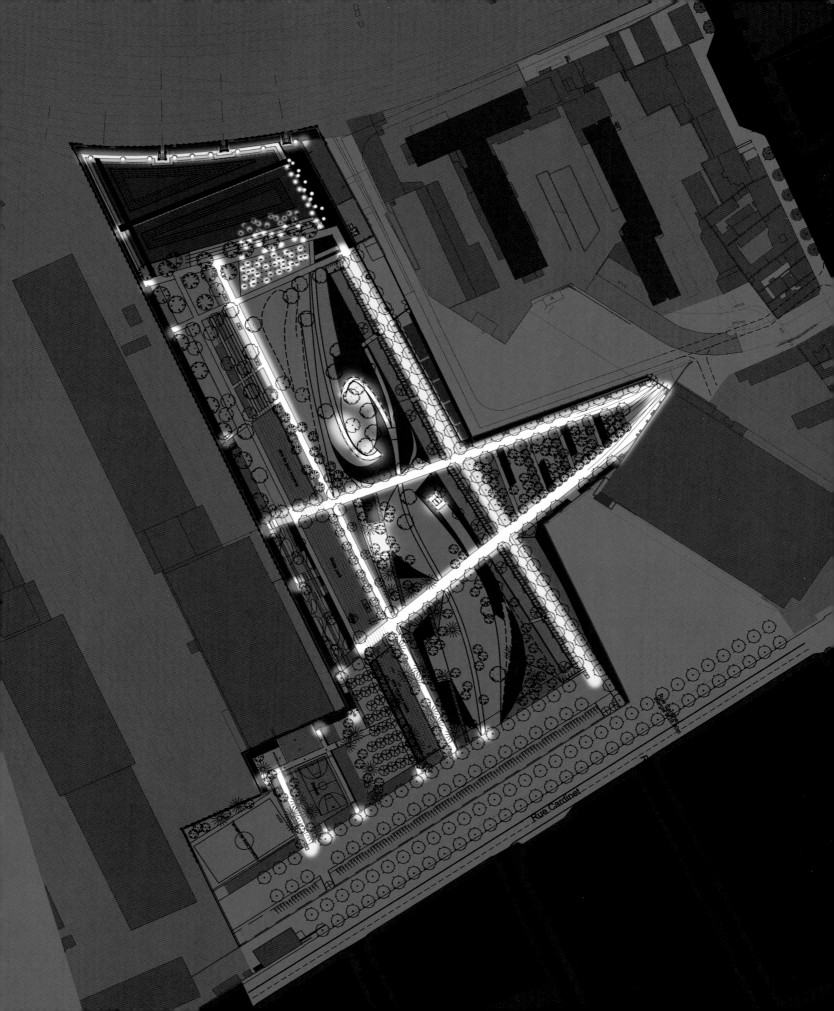

to the views and make the view and image of the park more attractive after dark. It is the manmade landscaping and volumes that are emphasised by the small button lights integrated into the amenities, turning them into new, virtual sketches.

The fountains, steps, pontoon and biotope basin in the north of the Park in front of the railway line are the subjects of another display, enhancing the theme of water at night. Small spots (devices with white LEDs) are randomly dotted around and in the fountains. Set into the ground, they are deployed up the steps and, finally, submerged in the water. This light "tableau", which is visible from the old Forge that will be turned into a restaurant during a future phase of the development, adds a delightful touch to the aquatic layout at night.

**Lighting the Paths**

The two main longitudinal paths that are extensions of the surrounding streets, and the two main transversal paths are lit using pairs of adjustable lights mounted on 5-metre posts on one side of the paths, in line with the trees. Minimum levels of lighting have been installed (an average of 10 lux is maintained) but the lights used on the ground nevertheless provide a sense of safety along the footpaths at night. The architectural lights (IP 65) are equipped with ceramic metal halide lamp 70W (3,000K, CRI >80). They can be fitted on site with a whole range of accessories (refractory glass to alter the distribution of light, coloured glass filter etc.).

The use of lights on posts is an excellent way of overcoming the problem of light pollution because the light is directed downwards, towards the ground. Moreover, it is flexible to use and allows for change and development, given the unusual location of the park within a district undergoing major alteration.

At the present time, the lights are directed onto the playgrounds and skate park but they can very easily be redirected into the trees if requested (once the trees have grown sufficiently), the lawns or a special event (music festival, temporary bandstand etc.).

### An Unusual Approach to Electricity Supply and Very Low Energy Consumption

The principle of centralised power connected to the grid has been used for the public lighting in the park and the peripheral lighting. Energy production from the photovoltaic panels mounted on the roof of the old Forge is injected into the grid and sold to the national power supply company before being bought back as conventional power, at low cost, as and when the need arises.

The lighting along the circle railway line, on the other hand, has been designed with a totally independent power supply (photovoltaic cells and an integrated storage battery). The average duration of light available for these devices is approximately 4 to 5 hours depending on the sunshine.

To further minimise the energy costs of providing the lighting, which are already very low for a 4.4 hectare park (9kW on full power), a number of different lighting programmes have been developed. The overall lighting is powered up at nightfall, when the park gates are closed (there are currently plans to open the park at night during certain periods of the year). From 10pm. in winter and midnight in summer, every second light will be lit to provide lighting along the paths and improve safety along the night-time perspectives. The nocturnal lighting displays are switched off at midnight. The lighting along the boundaries of the park can then remain on until daybreak.

巴黎克利西巴蒂诺尔公园的照明设计目标：
- 制造丰富而迷人的夜间氛围
- 提升主干道的安全性和保障性
- 阻隔从公园望向周边大型铁路地块的视线
- 美化公园的夜景形象，使其成为邻街区的城市景观

基地为巴黎申奥过程中的首选奥运村，照明设计定位基于可持续发展和高品质的环境质量。其他纳入考虑的因素：不妨碍动植物的生命，不增加巴黎的光污染。

### 阻隔公园尽端的视线

公园西接一片大型铁路地块，这一特殊的地理位置，使得在西边架设能够分割公园与远处无法到达区域的照明结构显得至关重要。次级园路路网的尽端沿着西边伸展，以小型白色灯具作为标识，这些灯具由在5米高的圆柱形金属桅杆上设置20瓦低能耗LED灯构成，远处即能看到。

### 照亮市内铁路环线

小型铁路"环"线在公园北角水池形成一个大弧形，根据巴黎"皇冠"照

明总体规划方案为其铺设照明,以6米高的木质灯柱上的蓝色灯光为特色,同时供电依靠独立的光伏电池。灯柱与铁轨的距离符合规定,沿岸而设,柱隙越来越宽,止于水池中央,进而增强了曲线效果。

每个灯具都配备了太阳能光伏板,灯柱里也嵌入了电池,由此保证了供电的完全自给。这意味着不再需要任何土木工程方面的工作(坑道、电缆护套、布线),除了混凝土地基。由此,灯具易于移动,这极大地便利了未来的改进工作和此区的重新规划。

**展现夜景**

公园各主题空间、季相变化、植物配置等均由景观设计师所设计,夜景各不相同。俯瞰公园的观察台设计了由1瓦LED灯构成的按钮灯具。这些精心设计所呈现的夜景效果,为公园的夜景增添了深度,使其更加迷人。这些小型的按钮灯具完全融合进娱乐设施里,突显了人造景观及其尺度,勾勒出一副新颖而真实的手绘图。

公园北部坐落在铁路线前方,喷泉、阶梯、码头、生物栖息池水的夜景则是另一番景致,增强了水在夜间的主题。白色LED小聚光灯随意散落在喷泉内外,有时嵌入地面或阶梯或水中。根据规划,这里未来将改建一个餐厅,届时从餐厅就可以看到这一灯的"话剧",为夜间水景布局增添了灵动的气息。

**照亮园路**

两条纵向主园路是附近街道的延伸,而两条横向主园路用成对的可调节灯照明,5米高的灯柱在园路一侧与树木成一直线。照明效果维持在最小强度范围内(平均10勒克斯),而步道地面的灯光则注重保证夜间的安全。

建筑照明(防护等级为65)安装70瓦的陶瓷金卤灯(高亮度3000K,CRI>80荧光粉)。这些灯具可以配套一系列装饰材料(调节光漫射的耐火玻璃,着色玻璃滤光片等)。

立柱灯具的使用克服了光污染,因为光线是向下洒,投射到地面的。同时,其也便于使用且适应各种变化和发展,比如公园区域特殊场地的重大变化。目前,这种灯具已经运用到运动场和溜冰场,同时如果有需求,它们也很容易移动并安装在树木里、草地上、特殊场合(音乐节、临时演奏台等)。

**供电和超低能耗的特殊方法**

公园内的公共场所和周边地带的照明采用集中供电的电网方式。屋顶上安装太阳能光伏板,所产生的能量归入电网并出售给国有供电公司,然后根据需求再以低成本回购传统电力。

铁路环线的照明自给供电(采用光伏电池和蓄电池)。这些灯具的平均照明时长大约为4到5小时,这取决于日照。

虽然对4.4公顷的公园而言,照明供电的成本已经很低了(照明全部开启的用电量为9千瓦),但是为了进一步将照明能源成本控制在最低,还是发展了各种不同的照明方案。傍晚,公园大门关闭后,所有的照明都会打开(现在有在一年中的特定时期在晚间开放公园的计划)。冬天晚上10点后,夏天午夜后,步道沿路的景观灯每两个将点亮一个,进而提高欣赏夜景的安全度。夜间照明将在午夜熄灭。公园边界沿途的照明将持续点亮至拂晓。

# Robert F. Kennedy Inspiration Park

## 背景灯光的视觉盛宴
——肯尼迪纪念公园

**Credits**
Location: Los Angeles, California, USA
Lighting Design: HLB Lighting Design (Tina Aghassian, Crystal Chen-Lim, Alexis Schlemer)
Landscape Architect: AHBE Landscape Architects
Design/Executive Architects: Gonzalez Goodale Architects
Historic Architects: Tetra-IBI Group, David Kaplan
Artists: May Sun, Richard Wyatt Jr. (Walls), Bobby Carlyle (Bronze Statue)
Water Feature Consultant: JOMA Design Studio

Representing sparkling constellation patterns from the night of Robert F. Kennedy's birth, tiny fiber optic luminaires embedded in blue glass pavement highlight a vibrant nighttime presence for the Robert F. Kennedy Inspiration Park located at the base of a 24-acre campus for the RFK Community Schools. Built on the site of the famed Ambassador Hotel where the Senator was assassinated, the one-third acre public park pays tribute to Kennedy's life by celebrating his social justice ideas and inspiring opportunities for discovery and reflection of these convictions by younger and future generations. HLB Lighting Design worked in tandem with a team of multiple architects and artists to fully realize the design of this special place within a dense urban neighborhood along busy Wilshire Boulevard. Design goals of integrating illumination into the landscape in order to establish an ambient light level, emphasizing focus on art installations and feature elements, honoring the historic nature of the site, led HLB to approach the lighting design with practical, yet creative techniques and solutions.

Lighting is concentrated on architectural and landscape elements giving focus to the artwork. The artwork consists of a 110 foot wide sandstone wall featuring Senator Kennedy's portrait and engraved with inspiring quotations from Kennedy and other advocates of social justice, as well as a parallel 24 foot wide stainless steel wall with a cut-out symbolizing a ripple in the water connoting Kennedy's "Ripples of Hope" speech. The sandstone wall is illuminated by shielded, ingrade metal halide uplights providing focused illumination and minimizing light pollution. The ambient glow provided, does not compete with, and allows for the constellation patterns to shine. Layering of light is achieved via ingrade linear fluorescent sources which accentuate the expression of the ripple in the stainless steel wall through to the sandstone artwork beyond. The surrounding fiber optic luminaires fulfills the artists' vision and "twinkles" slightly via a sparkle wheel creating a sense of movement and dynamism. Energy-efficiency and easily maintainable luminaires with high color rendition were important considerations in the design process.

As visitors travel westward, the art installation transitions into adjacent walls that terminate at the hotel¡'s original historic sign pylon and a recreated

bronze sculpture and fountain. To illuminate these iconic elements, HLB utilized halogen uplights set in a ring formation inside the fountain, lighting both elements and bringing a three-dimensionality to the space. The illumination serves to dramatically silhouette the dark bronze sculpture against the contrasting white pylon, generating an exciting interplay of shadow and light.

Groves of trees which will eventually grow into a spectacular "green screen" are outlined against walls with the same shielded, ingrade metal halide uplights for continuity. Nearest to the roadway, illumination is supplemented by step lights for security which contain a glowing face to provide more light on occupant's faces versus a louvered fixture. The team chose to not illuminate the trees themselves in order to minimize light pollution and to emphasize the tree profiles.

Throughout the park, low level compact fluorescent steplights and bollards are integrated into walls and steps supplying safety lighting and a sense of comfort to visitors while also avoiding obstructions to the view - allowing for inspiration, contemplation, and possibly, a ripple of hope.

嵌入蓝色玻璃人行道的微小光纤灯为坐落于RFK社区学校9.7公顷校区之上的肯尼迪纪念公园呈现了充满活力的夜晚，重现了罗伯特·费朗西斯·肯尼迪生出之夜的群星荟萃。

建于当年参议员被暗杀的著名的大使饭店遗址之上，公园通过颂扬他的社会正义理念和激发年轻人与后代们发现与反映这些信仰的机会，向肯尼迪的一生致敬。HLB灯光设计公司与众多设计师和艺术家团队合作，充分实现了繁忙的威尔夏大道密集的街区中这个特殊地方的设计。设计的目标是将照明整合到景观中以建立环境照明水平，强调对艺术装置与特征元素的专注，尊重场地悠久的历史，这些目标使HLB的照明设计技术与方案更为实用而不失创新。

照明集中于建筑和景观元素，来突显艺术作品。艺术作品包括：宽33.5米、主体为肯尼迪参议员的肖像并刻有肯尼迪及其他社会正义拥护者的语录的砂岩墙，和平行的宽7.3米，象征肯尼迪"希望的涟漪"演讲的不锈钢墙。金属卤化物射灯不仅照亮了砂岩墙，还减少了光污染。周围的微小光纤在光轮上轻轻地闪烁着，营造出动态的感觉。节能和便于维护的高演色性照明灯是设计过程中要考虑的重要问题。

当参观者向西前行时，艺术装置变成邻近大使酒店原历史性标志塔的墙，和重造的青铜雕塑与喷泉。为了照亮这些标志性元素，HLB利用卤素射

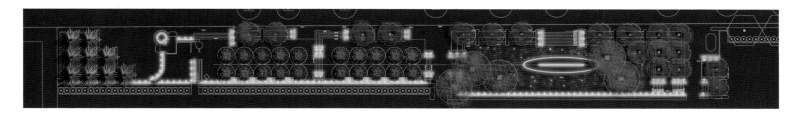

灯在喷泉里围成一个圈，从而点亮所有元素，形成一个立体的三维空间。照明使深色的青铜雕塑与白色的电缆塔相比，更具有引人注目的轮廓，产生一种光影互动的惊奇效果。

在有一排连续金卤灯的墙边，将长成一片壮观的"绿幕"的树木理出轮廓。在最靠近道路的地方，保证安全的照明由踏板灯补充。设计团队选择不照亮树木，是为了减少光污染和强调树木的外观。

公园各处，低电平的紧凑型荧光踏板灯和系缆桩在墙与台阶处提供安全照明，带给游客舒适感，同时也避免妨碍景色——出于对灵感、沉思、以及或许还有，希望的涟漪的考虑。

# Wolfsburg Automobile City

光影交错
——沃尔夫斯堡汽车城

**Credits**
Location: Wolfsburg, Germany
Landscape Architect: WES & Partners
Cooperation Team: Office Max Wehberg, Hamburg

It is the Automobile City's mission to enter into a new, open dialogue with customers and visitors. In the Car City the company presents its most basic values while also displaying its products. It portrays its past and provides an insight into its know-how; at the same time the city receives its visitors as though they were its own private guests. The Automobile City is not only a presentation of the corporation itself. The new manufacturing approach—no noise pollution and smoke as experienced in the past—also gives the city new opportunities. Not only does this strengthen the dialogue between the company and its clients, but it also promotes a dialogue between the city, its inhabitants and Automobile City. The Mittellandkanal, which separates the city and the production plant, is bridged, and the banks can be accessed, creating a

new centre of focus.
A new kind of park has been developed, which provides the visitors—in addition to the automobile themes presented in a wide range of different buildings—with a differentiated outdoor space, of which someone once said that it is the soul of the car city.
Development and monitoring of temporary events "Winterspiele" (Winter Games) and "Frühlingserwachen" (Spring Awakening) held at the park.
Since 2000, different activities have been taking place over a period of three months respectively.

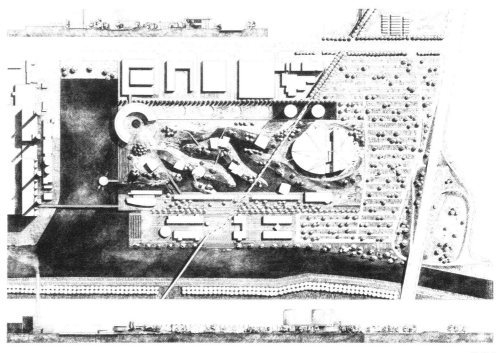

景观设计的目的是要促进汽车城同客户和游客间自由坦诚的交流。公司不仅通过汽车城展示其优质的汽车产品，更展示了其在城市中举足轻重的影响和作用。人们在这里不仅能领略到公司的历史、技术和产品，更能在这处舒适的休闲空间内感受到宾至如归的亲切感受。汽车城不仅仅代表着公司的文化和风格，更以其新型的生产方式——无噪音无烟尘，展现了城市的新风貌。这不仅促进了公司同客户间的沟通和联系，更加强了城市、居民及汽车城间的和谐关系。穿过城市和生产区的中德运河上还架起了一座桥梁，同两岸的风景一起构成了迷人的休闲空间。

汽车城中建造了一座新概念公园，向游客提供了从各种不同建筑中表现出来的汽车主题和一个与众不同、曾被人称为汽车城灵魂的户外空间。

自2000年以来，公园中举办了各种不同的活动，如冬季运动会和迎春活动等。这里已经逐渐成为城市中一处充满魅力的特色空间。

# An Eventful Path

足下的艺术
——悉尼奥林匹克公园历史大道

**Credits**
Location: Sydney Olympic Park, NSW, Australia
Client: Sydney Olympic Park Authority
Landscape Architects: Aspect Landscape Architecture & Urban Design
Collaborator: Feeder Associates (Graphic Design)

Situated somewhere between Landscape, Architecture + Design, the eventful path is a project by Aspect Sydney Landscape Architecture in collaboration with Feeder Associates Graphic Designers. It is a public installation which celebrates and commemorates Australia's major international cultural and sporting events held at Sydney Olympic Park.

The installation is a 45m long path comprising blocks of cast coloured glass, stainless steel channel, fluorescent lights and honed concrete pavement. The linear path is an abstract time line, which celebrates the past and looks to the future. Installed as a complete work, it evolves over time with the inscription of bronze plaques to commemorate each new major event held at the Sydney Olympic Park. Blank plaques will be melted down and re-cast with new inscriptions when a significant event has been held. It is the intent of the installation that the laying of each new plaque establishes an ongoing ritual at Sydney Olympic Park.

Lit at night, the installation brings joy to the public domain and illuminates the way to the parks major event precincts. The design was generated with the idea of not producing more "art objects" in the landscape. The result is a beautifully detailed in-ground installation which brings colour to the public domain and animates the notion of memorial. Situated at the forecourt to the Sydney Olympic Park station, the Eventful Path is a public project that can be viewed anytime. The lights are programmed to come on each day at 5.30 pm.

The use of coloured glass in the ground is not common in landscape architecture and was the biggest challenge of the project. The custom made coloured glass is cast with a textured top face to

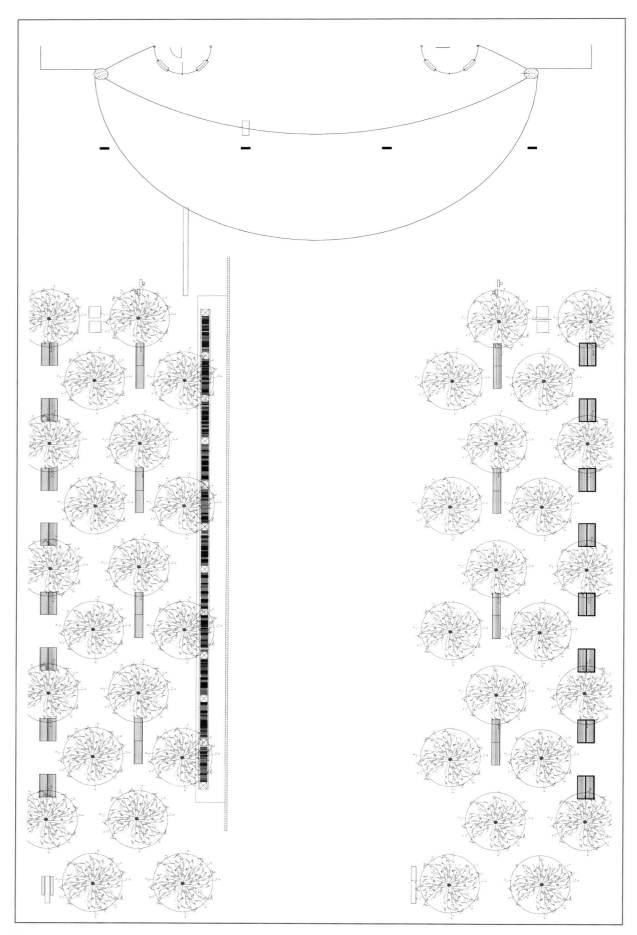

create a slip-resistant surface. The installation is also designed to withstand large amounts of pedestrian traffic (during major sporting events) and occasional heavy vehicular traffic (fire trucks etc.) without cracking. Due to vast amounts of pedestrian flows in and out of the train station during big events, the builders had to ensure that all edges were flush to the surrounding pavement to a tolerance of +/-3mm. During the day, the installation appears seamlessly integrated into the station forecourt. At night, it turns into a beautiful, glowing line of colours and people are attracted to walk along its length, reading the inscribed plaques as they travel.

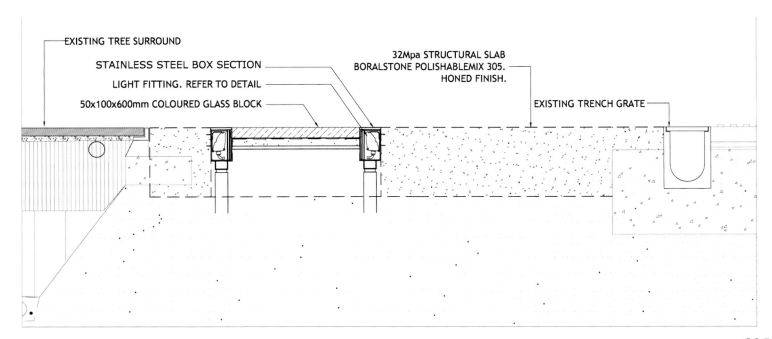

奥斯派克景观/规划设计公司悉尼分公司与Feeder平面设计公司携手合作，设计了一条色彩斑斓的发光的人行道。这条大道的设计类型介于景观与建筑设计之间，薄金属板上记载着在悉尼奥林匹克公园举行的重大国际文化体育事件，使得一幅幅的历史画面重新在人们眼前展开。

这条历史大道长45米，由一块块长条形的彩色玻璃、不锈钢、荧光灯和磨沙混凝土地面组成。道路两边用不锈钢长边围合，中间用各种大小相同的彩色玻璃平行排列。

这条直线形的道路象征的是一个纪念过去，展望未来的抽象的时间列表。往后在悉尼奥林匹克公园举行的任何重大文化体育盛事都将被记录在大道上预留的青铜薄板上。但凡有重大赛事举办，青铜薄板将会被重新熔解，铸刻上新的描述。这样随着时间的推移，历史大道上记录的重大盛事也会逐年增多。悉尼奥委会希望每一块新铸字青铜板的产生都能在悉尼奥林匹克公园内举办一个庆典仪式来予以庆祝。

坐落在奥林匹克公园火车站前庭的历史大道是一个可随时供人参观的公共景观。每当夜幕降临，五彩玻璃下的灯光就会被点亮，历史大道那迷人的彩色光芒总能吸引无数游人新奇的目光，同时也照亮了通往奥林匹克公园主场地的道路。设计

所遵循的理念是无须在这片景观带上再添加更多的艺术形式，所以通过设计师巧妙的构思，这条精美的、造型独特的纪念道为这片公共空间带来了活泼的色彩，也通过新颖生动的形式记录下了各重大赛事。

将彩色的玻璃用于地面铺装的手法在景观建筑中是比较少见的，所以这便是设计者在此项目中遇到的最大的挑战。彩色玻璃的上部被铸上了一层粗糙的表面，以达到防滑的效果。同时，这条玻璃道也必须要承受由于大量人流的踩踏和重型交通工具经过时(例如消防车)所带来的压力，而不会破裂。考虑到在重大赛事召开时进出车站的大量人流，建造者必须确保每块彩色玻璃与周围的路面保持在同一水平面，允许的高度差不能超过+/-3毫米。

在白天未打开底部灯光的时候，历史大道协调地融入周围的景观之中，并未展现出其独特的设计美感。然而每到夜间，这条发光的彩色大道总能吸引人们驻足观赏，阅读刻在青铜薄板上的赛事简介，感受体育赛事所带来的激情和力量。

# Glostrup Town Hall Park

## 星空下的童话王国
## ——Glostrup市镇厅公园

**Credits**
Location: Rådhusparken, Nyvej, Glostrup, Denmark
Scale: 5,500 m²
Client: Municipality of Glostrup
Lighting Design: SLA
Team: Stig L. Andersson, Hanne Bruun Møller, Stine Poulsen, Christian Restorff-Liliegren
Cost of construction: 1.2 Mill. € (Nyvej 1.7 Mill. €)

This project in Glostrup has not only changed the town physically: it has caused the residents to change their behaviour. The residents in the area say that they re-route their journey to work through the Town Hall Park and often choose to stay a while to enjoy the sensory pause. Pollarded lime trees from a previous road have been preserved as a contrast to the new interventions, supplemented with trees otherwise found in the surrounding gardens. In this way a connection is created between the public space of the Town Hall Park and the residents' private gardens, thus establishing intimacy. The Town Hall Park is constructed as a square in the centre of Glostrup's modernist suburban space, and is the last link in a project to connect the central parts of the town between the station, church, town hall and shopping centre. To create experiences in the heterogeneous built environment, the square has been given a prominent character, i.a. by the surfacing of Norwegian "otta pilarguri" slates, that abut the facades of the surrounding buildings. The surfacing is contoured so the visitor becomes aware of his or her own body and balance when he/she moves through the park. On Nyvej lime trees have been planted in double rows that in time will transform the road into a classical avenue. The light is in the surfacing as on a runway and separates cyclists and pedestrians.

位于丹麦Glostrup的这个项目不仅在空间结构上对城镇进行了改造，同时还引导着居民行为的改变。自从公园建成开放，这里的居民会在家和公司间的路途中短暂地停留，享受空间的魅力。之前道路边的树木得到了保护，如今的它们修剪整齐，与周围的新环境形成了强烈的对比。周围花园中的其他植物则与这些树木一起构筑了绿色环境。设计师以这样的方式营造了公园的开放空间与周围居民私家花园的联系和互动。公园建设得如同环绕在城郊中心的一座广场，并将中心城区同车站、教堂、市政厅和购物中心等地方连接起来。为了营造这处环绕于不同建筑环境中的广场，设计师特地采用了挪威石板作为铺地，与周围建筑的外立面形成了呼应。地面的铺设顺应了原有的地形，游客在行走的过程中能够感受到空间的平衡。树木采用对植的方式，把道路变化为古典的林荫大道。地面上的灯光将机动车道和人行道等不同的道路进行了分隔。

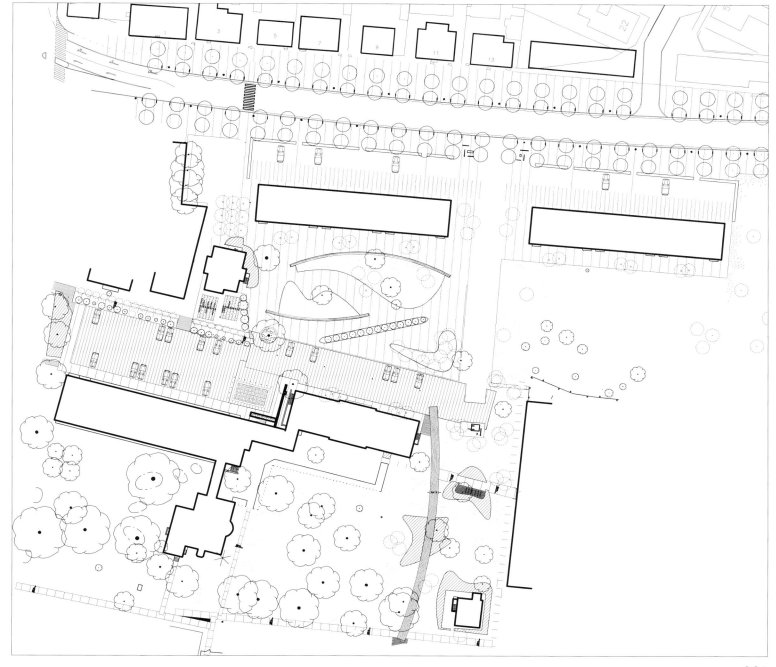

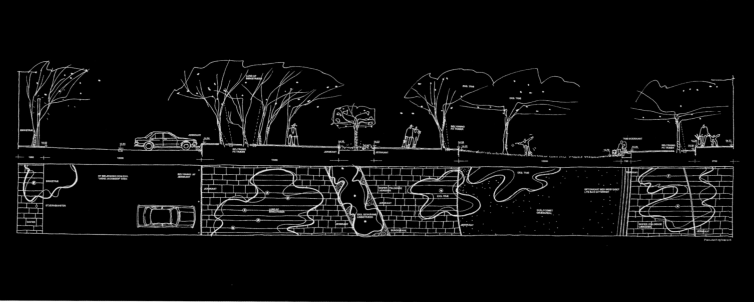

# The Garonne in Toulouse

新技术演绎的生态照明
——图卢兹市加伦河

**Credits**
Location: Toulouse, France
Project Owner: Toulouse City Council
Lighting Design: Roger Narboni, CONCEPTO agency - Sara Castagné, Mélina Votadoro, project leaders
Technical Design Office: Beture Infrastructure, Jean François Glière, Project Leader, Georges Toussaint, Assistant
PR: MC3, Mireille Cerboni
Lighting Devices: Targetti Extérieur Vert
Installation: AMEC SPIE
Civil Engineering: SPIE Batignolles

The River Garonne, which flows through Toulouse city centre in a long graceful curve, was one of the main themes given prominence in the Lighting Master Plan for the city, discussed and approved in February 2004. The lower and upper quaysides along the banks, the riverside trees, the bridges, the "Prairie des Filtres" Park, the tip of the Ramier island, and the buildings and churches overlooking the waterway were all gradually floodlit, creating an extensive nighttime riverscape over a distance of more than 1.5 kilometres.

The lighting of the Garonne itself, with the lights set into the river bed and stretching from one bank to the other at the causeway known as Chaussée du Bazacle, is the most outstanding feature of this innovative, ambitious project designed to give Toulouse an unusual, enchanting point of interest after dark.

The ford called Le Bazacle, a shallow passage over hard rock in the river bed, was for many years the only crossing-point on the Garonne and this undoubtedly explains why Toulouse was founded here. The dykes built in the 12th century to hold back floods were interconnected in 1248, forming the causeway. Today's concrete construction, which belongs to the French electricity company (EDF), zigzags between a hydro-electrical plant and the front of the La Grave hospice. The causeway consists of two, stepped sections, one gently-sloping and only lightly-structured and the other tall and vertical, and it takes on a very different appearance depending on the season and the rate of flow. In summer, a mere trickle of water dribbles over it; in winter, the water surges over the construction, concealing it beneath a rushing, foaming waterfall.

This "broken line" connecting the two river banks, though a symbol of Toulouse, is scarcely visible at night. It was decided to enhance it using a partially dotted, coloured pathway of light.

The 247-metre track of light is created by a discontinuous row of 265 identical lights 80 cm long (with a protection grading of IP 68), set into a concrete construction built on site to follow the line of the existing causeway.

Because of the characteristics of the installation and the very strong constraints inherent to this submersible site, special lighting equipment was developed by Roger Narboni, working jointly with Targetti Extérieur Vert, a manufacturer specialising in products with a high protection grading. Given the requirement for low voltage electricity passing through an underwater construction, LED technology was the obvious choice but in this instance the diodes were used to produce high-output optical conduction, a system that is particularly unusual and innovative.

Each device includes a translucent bar with lateral emission and an aluminium reflector. It is lit at the ends by two 1W cyan LEDs (wavelength between 495 and 510 nanometres) that produce very intense light. The LEDs are long-lasting, with a life of more

lights are set in line edge-to-edge and side-by-side in groups of six at 80cm intervals. They consist of a stainless steel profile capable of withstanding difficult external conditions (prolonged submersion in fresh water), covered with thick, totally flat polymer resin (60-joule shock resistance).
The low voltage electricity supply cables are sheathed and pass under the lighting inside the concrete

later date, the disposable lighting equipment can be removed and replaced with identical equipment without any need for further work on the concrete construction.
In the darkness, the lights create a dotted line that is clearly visible in the nighttime riverscape. The coloured line changes naturally depending on the Garonne's rate of flow. It is sharp in summer but

becomes more and more blurred during the winter as the current and flow rate of the river changes.

The site was accessible only in the summer months when the water level was at its lowest, and the project required complex work in the river and appropriate safety measures (floating pontoon to access the work site, cofferdams to protect the construction work, a sensor to monitor any rise in the water level, a hooter to warn of impending danger, with a boat always on hand etc.).

The light display is part of an overall ecological approach to projects. The extensive nighttime display uses approximately 900W. The equipment does not produce any heat and does not disturb the river's natural environment, affecting neither plant life nor animals. The concrete construction was built with particular care to ensure that no building materials were lost in the river bed.

加伦河流经图卢兹市，形成长长的优美曲线，成为2004年城市照明总体规划里的主要亮点，经过讨论并通过审批。河岸的高地和低地码头区、滨河树木、桥梁、雷费尔特草地公园、拉米尔岛屿的尖部、大厦、教堂，逐渐装置泛光灯，在光影中俯瞰着加伦河，营造出绵延1.5公里的大面积滨河夜景。

加伦河的照明设计集中在连接着两岸的堤道河床的灯具嵌入，这是此次独具创新、富有远见的项目里最为杰出的特色，为图卢兹市的夜景注入迷人而非同一般的元素。

堤道是河床里硬石上的浅滩道，曾是数年来加伦河的唯一渡口，这也毫无疑问地解释了图卢兹市发源于此的原因。

堤坝建于12世纪，用于防洪，于1248年相互连接，形成了堤道。今天的混凝土结构属于法国电力公司，蜿蜒在一家水电厂和拉格雷救济院前方之间。堤道由两个阶梯状部分组成，一个坡度较缓、略有架构，另一个高高地竖立。堤道基于不同的季节和水流速度而呈现不同的样子。夏季，细小的水流从堤道滴下，冬季，河水从混凝土结构上汹涌而过，将堤道隐藏在湍急的、泡沫飞溅的瀑布里。

"断线"连接着河流的两岸，虽然是图卢兹市的一个象征标志，却在夜间极少显现。为此决定采用虚线式的彩光道来提升其形象。

256条相同的80厘米长的灯具(防护等级IP68)以断续的方式铺装出一条247米长的照明道路，嵌入基地上铺筑的混凝土结构，勾勒出现存的堤道。

根据这一装置的特点及其沉水基地固有的局限性，罗杰·那波尼和一家专门生产高防护等级产

品的制造商"Targetti Extérieur Vert"共同合作，特别研发了照明设备。即便水下结构对低压电有需求，LED技术本是当然之选，但是此处却采用了特殊的创新型二极管照明系统，用于生产高输出光传导。

每个照明设备都装有一个横向发光的半透明条管和一个铝反射镜。灯具两端由两个1瓦青色LED灯（波长495～510纳米）点亮，形成的光源十分强烈。LED灯可以持续照射很久，寿命超过50 000小时。照明设备周围的光源可见区域为10厘米宽，80厘米长。灯具一个连一个，边对边地六个并为一组，每组的间距为80厘米。它们的侧边以不锈钢围合，能够抵御艰难的外部条件（延长了灯具在淡水中的寿命），顶部则覆盖厚厚的、平坦的高分子树脂（抗冲击性能为60焦耳）。

低压供电电缆装在电缆护套里，铺设在照明设备下方，嵌入基地上铺筑的混凝土结构里。变压器设在200米远的地方。如果日后出现断层和塌

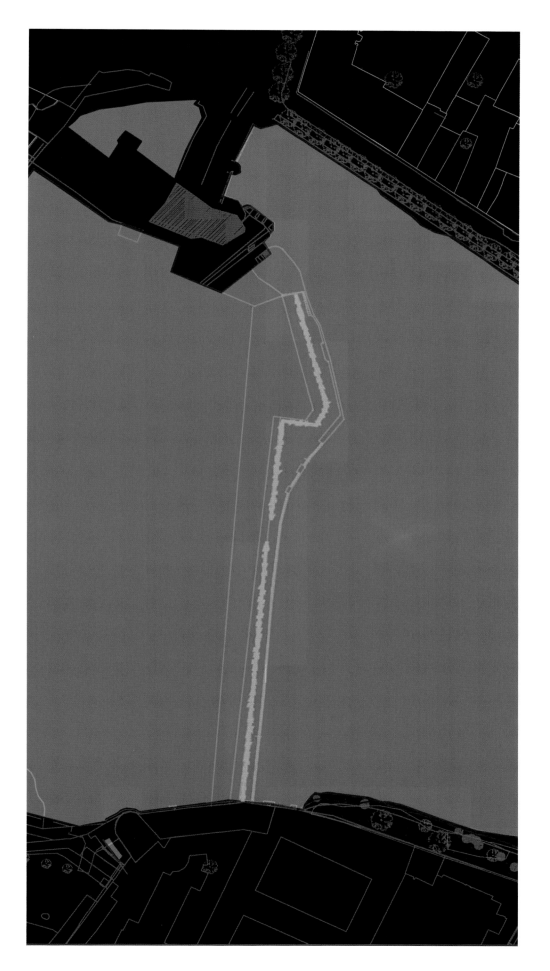

陷,可以将这些用后即丢弃的照明设备移除,换之以相同的设备,无需对混凝土结构做进一步的作业。

夜色中,灯光营造出一条虚线,在滨水夜景里清晰可辨。彩色线条的自然变化取决于加伦河的水流速度。夏季,它轮廓分明;冬季,随着河水流量和流速的变化则越来越模糊。

基地的可达性仅限于夏季月份,而且是水位在最低点时,完成了项目要求的复杂工程,也采取了适当的安全措施(通往施工现场的浮桥,保护施工的围堰,监测水位上涨的传感器,危险时刻的警笛,随时可达的船只等)。

灯光演绎是整个工程生态手法的一部分。大面积的夜景照明所耗电能近900瓦。照明设备并不产生热能,也不破坏河流的自然环境,对动植物等都没造成影响。混凝土结构施工过程中,特别关注并确保了建筑材料没有遗失在河床里。

# Slater Mill Falls

白色光照的梦幻效果
——斯莱特米尔瀑布

**Credits**
Location: Pawtucket, RI, USA
Lighting Design: Abernathy Lighting Design
Photography: Abernathy Lighting Design Inc.

In 2004 Abernathy Lighting Design was asked to highlight the waterfalls of the Blackstone River located outside Slater Mill in Pawtucket RI. When approached about the project it was mentioned that the client wanted the waterfalls to be lit as brilliant as Niagara Falls. The waterfalls played an important part in Rhode Island's history, serving as the power for the first water powered cotton spinning mill in North America. It was only appropriate that the falls sparkle with the history it represents.

It was important that the mills be seen and the fixtures not. The first point of action was the question of where the team going was going to mount the luminaires. Being an outside venue there weren't very many options that allow for maintenance ease as well as power. Upon first examination, the initial point of placement was under the Blackstone Bridge which runs along the river. It would have allowed the light to be seen on falls without seeing the luminaire itself. But upon learning of the condition of the bridge as well as the cost to run power out to the location, not to mention that this river has been known to flood, meaning the lights would be submerged at that time, so an alternate solution had to be found. The next best location turned out to be along the stonewall that surrounded the perimeter of

the river. This location allowed the luminaires to be easily maintained while still allowing for a good angle to illuminate the falls.

However with the mounting position being approximately 110m away, the team was now faced with a distribution challenge. How were they to make the falls sparkle and not let any light pass into the neighborhoods? Whatever luminaire that was chosen had to have a beam spread as tight as possible and be very powerful to carry the light the necessary distance.

When it came to power there was no doubt that Metal Halide was the source to do the job. It was the

most powerful reliable lamp and we were able to put it in a reflector that would give us a 4 deg. beam. It took four luminaires to cover the entire falls at the 110m distance away. The luminaires were mounted next to the stone wall and aimed appropriately. The City unfortunately was unable to conceal the luminaires in an outdoor decorated enclosure due to the lack of funding, but did make the decision to install the luminaires and light up the falls. As time has passed since the initial installation they still wish to obtain this enclosure but have found that the luminaires have such a precise distribution that no one notices the fittings even with their size!

These luminaires not only light up the falls but grazed across the rapids created by the water at the bottom of the falls. This enhances the dramatic effect as the light enhances the strength and power of the water. It was originally thought that colors would be nice shinning on the falls, but an onsite mock-up concluded that in this instance the pure natural white light made the biggest, dramatic and most beautiful impact.

As stated early there is a great deal of history related to this particular water fall and it is now celebrated each night with the precise lighting design. As part of a several year plan the City of Pawtucket is looking to upgrade its image and one way is by enhancing the quality of light during the evening hours. Abernathy Lighting Design has worked on several master plans along the Blackstone River and Slater Mills Falls for the City of Pawtucket and will continue to work toward making sure it employs the most energy efficient, maintenance friendly but most importantly quality lit exterior places.

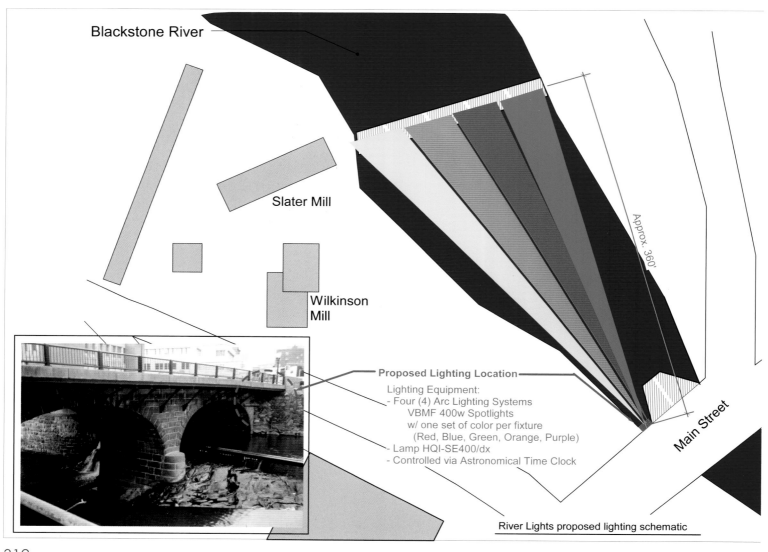

River Lights proposed lighting schematic

2004年，阿伯纳斯照明设计所受邀为斯拉特工厂外围的黑石河瀑布进行加亮设计，该工厂位于罗德岛州的波塔基特。工厂要求将黑石河瀑布塑造成如同尼亚加拉瀑布那样华丽。作为北美第一座水力棉纺工厂的一部分，瀑布在罗德岛历史中举足轻重。因此将瀑布与其象征的历史融合是最为贴切的做法。

看得见的厂区和看不见的照明设备是两处重点。规划的首要在于团队选择安置照明设备的地点。由于在室外，兼顾灯具维护和电力的方案并不多。在第一次试验中，首选的地点在沿河黑石桥底下。如此安排的话，可在瀑布中看见灯光，却看不见灯具。但考虑到桥的状况，及从此发电的成本，以及河水泛涨时灯具将淹没于水下等情况，最终形成了一套替代方案，将灯具沿着河流周围的石墙安装，如此安排可使灯具易于维护，且仍有良好的角度照亮瀑布。

然而，由于支点位置远在110米开外，团队面临着灯光分配的难题。如何使光线照亮瀑布，且不影响周围邻里？选择的照明灯具的光束发散度必须极为紧密，并能将光线在规定距离内有力地映射在目标物体上。

在实际运用中，金属卤素灯无疑是首选。它是最值得信赖的灯具，我们将其安装在反射器上，可以打出4度的光束角。从110米开外的距离覆盖整条瀑布需要4盏灯。灯具安装在石墙的旁边，照射目标十分明确。不幸的是，由于资金原因，波塔基特市政府无法用室外装饰围栏将灯具隐藏起来，但还是决定在瀑布上方安装灯具，照亮瀑布。随着最初安装后时间的流逝，他们仍然希望能装上围栏，却发现照明的公布如此精确，以至于就算设备的尺寸不小也没有引起任何人的注意。

这些灯具不仅照亮了瀑布，而且和瀑布底下湍急的水流纵横交错。灯光烘托着水的力量，催生了戏剧效果。起初认为有色彩在瀑布上闪烁将会很好，但从现场实体模型得出的结论却是，在这种情况下纯粹、自然的白色光才能制造出最强而有戏剧性和最美丽的效果。

如同前面提到的，这处特别的瀑布蕴藏着鲜活的历史，这段历史在每晚灯光的照明下得到映衬。作为年度规划的一部分，波塔基特城致力于提升自身形象，方法之一便是提高夜晚的照明质量。阿伯纳斯照明设计所已经为波塔基特城的黑石河及斯莱特工厂瀑布沿岸进行了数项宏伟的规划，并为确保使用最具能效、最容易维护且最具存在感的照明设计而继续努力。

# Shed 1, Princess Wharf

## 悬浮的"交流屏"
### ——公主码头1号棚

**Credits**
Location: Hobart, Tasmania, Australia
Scale: Interior 4,500 m² approx; site 7,500 m² approx
Lighting Design: PointOfView
Architect: Circa Architecture / Robert Morris Nunn
Landscape/urban
Designer: Spackman Mossop Michaels
Photography: David Becker, We-Ef
Awards: IES VIC Lighting Design Award of Excellence 2011

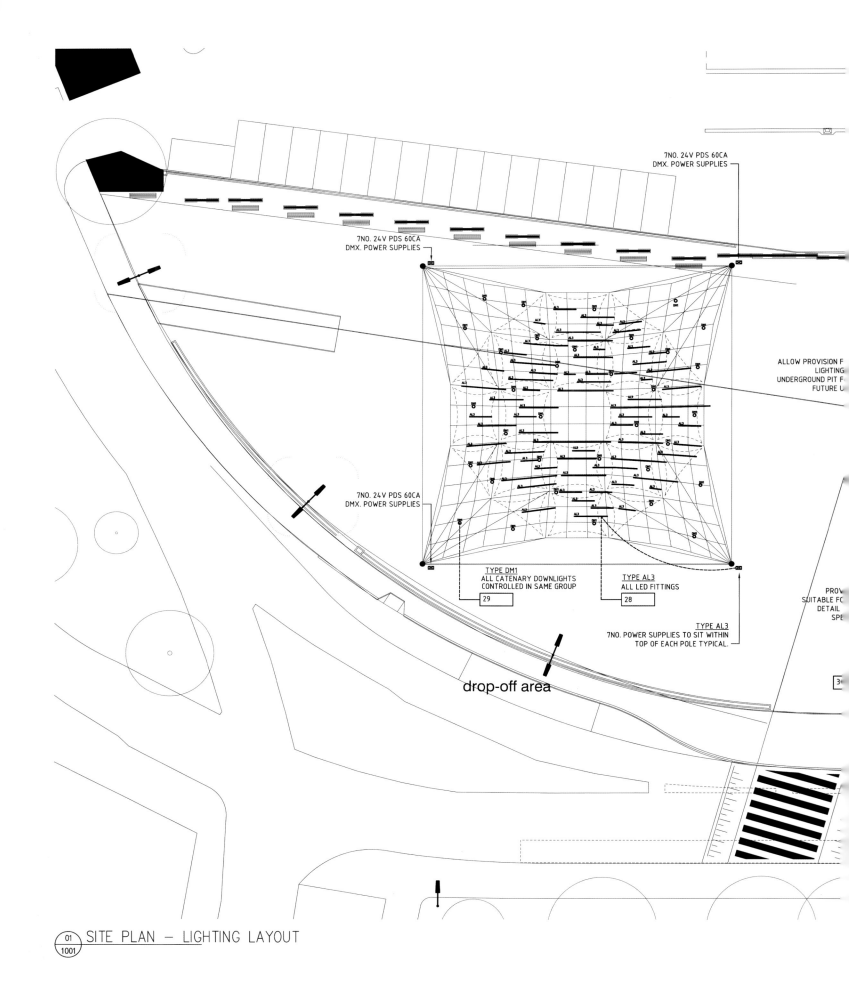

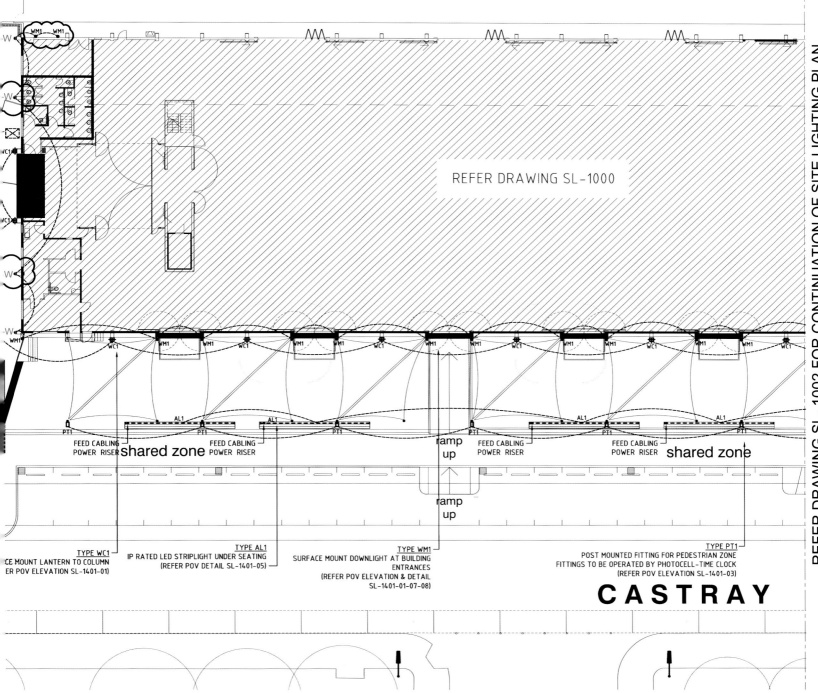

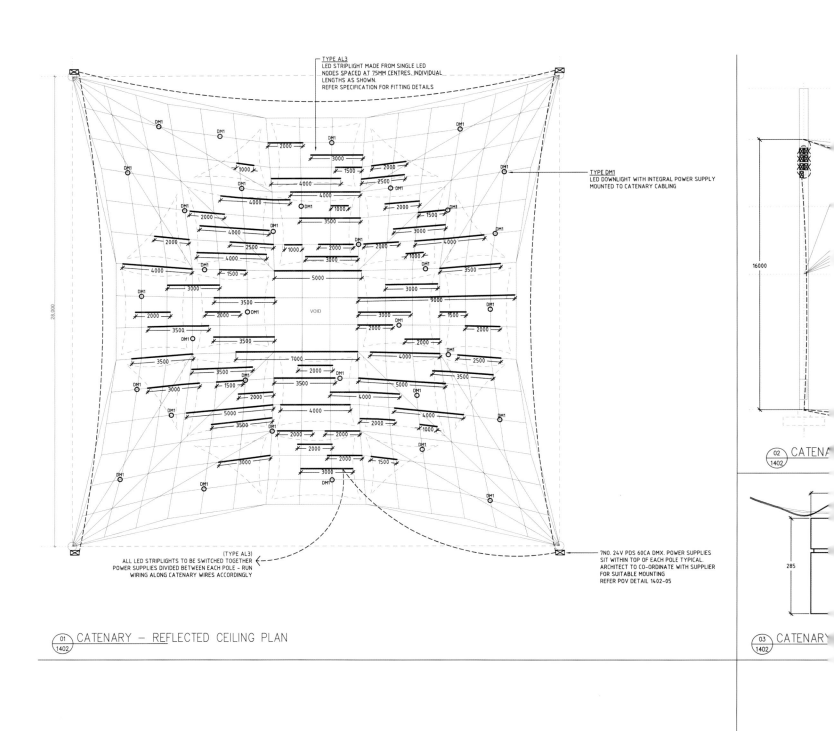

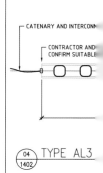

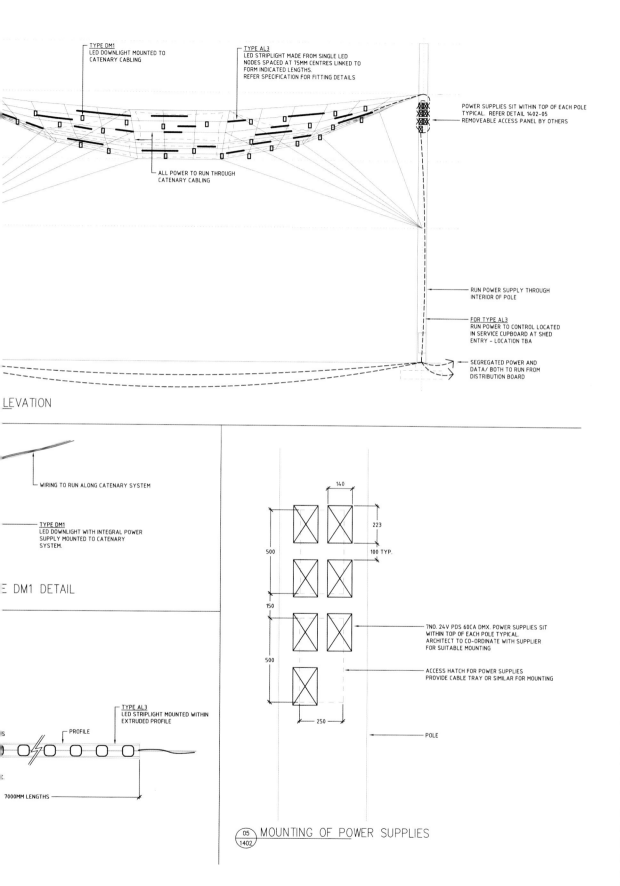

Originally constructed as a warehouse for Hobart's Port Authority, the governing principle of the Shed 1 project was to transform a crude utilitarian building into a multi-purpose hall as a public facility, and compliment the popular Salamanca area adjacent to the Princess Wharf. The building is approximately 130m long by 35m wide. The brief stipulated that the building should serve a broad range of public needs, from hosting the annual "Taste of Tasmania" food festival, arts events, concerts, exhibitions, rehearsals & large meetings. A large exterior performance space was required at the forecourt for theatre and other live events. The architectural intent was to preserve the "honesty" and simple construction of the original building, a sentiment which PointOfView embraced through a pure and essential approach to lighting design both inside and out.

The essence of the lighting scheme is to accentuate the building as a backdrop for the various functional demands, and also as a feature element to the local environment at night. By accentuating the simple materials and structural elements, a strong spatial narrative is created which is grounded in the building's history and context.

### Exterior

Shed 1 is a large building seen principally against the backdrop of the open sea. The darkness of the harbour and relatively low ambient conditions elsewhere means that very little light is required to provide a strong, yet balanced effect. The façade scheme comprises two key lighting approaches; the use of lanterns—soft light for exterior functions on the new deck area, and dramatic shafts of light that graze the structural columns of the building.

A shared pedestrian & cycleway is lit by luminaires mounted to the awning structure, with careful consideration to the influence these lights may have on the overall drama of the building. The forecourt has been designed to serve public performance, and features a large shade canopy supported by 4 masts each 16 metres tall. This structure provides the method for general area lighting (via suspended LED cans), but also works as a piece

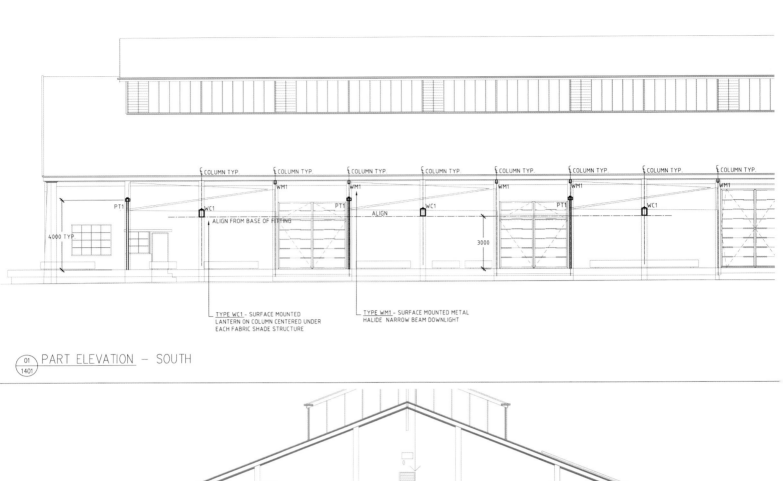

**01 / 1401** PART ELEVATION – SOUTH

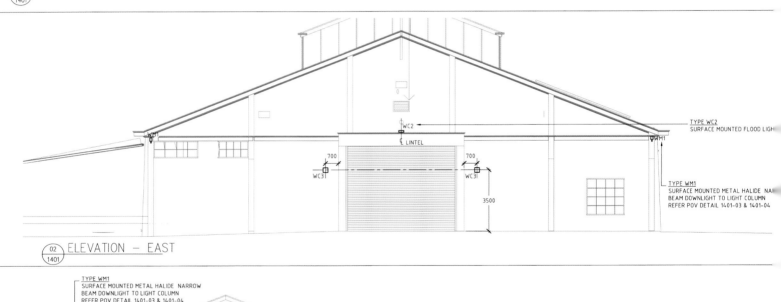

**02 / 1401** ELEVATION – EAST

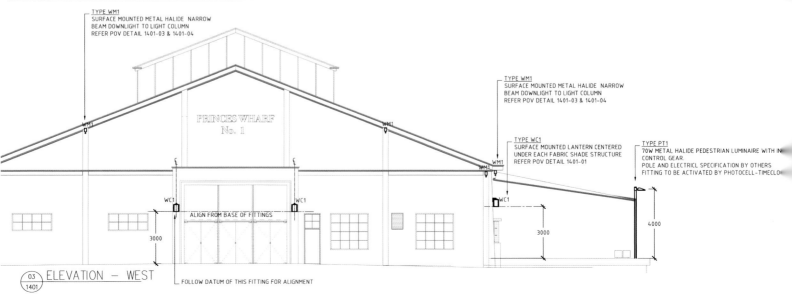

**03 / 1401** ELEVATION – WEST

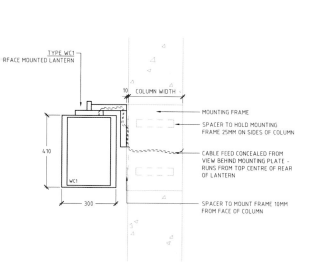

04/1401 TYPE WC1 — LANTERN ELEVATION

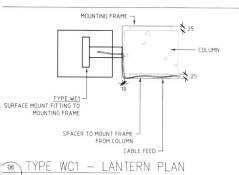

06/1401 TYPE WC1 — LANTERN PLAN

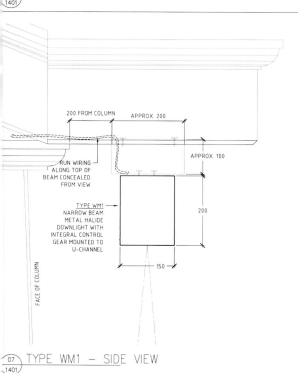

07/1401 TYPE WM1 — SIDE VIEW

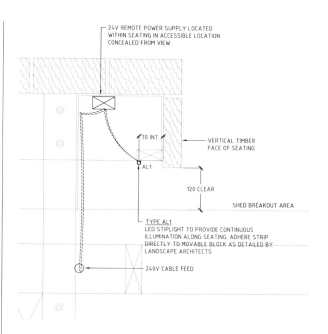

05/1401 TYPE AL1 — LED TO DECK STAIR

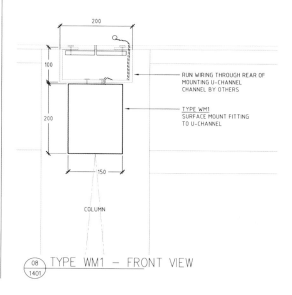

08/1401 TYPE WM1 — FRONT VIEW

of public art; a programmable kinetic LED canvas integrated with an audio system forms the illumination infrastructure intended for use by local media artists. Artists are commissioned on a term basis; effectively they have a programmable nodal RGB lighting system with audio outlets with which to create their art-piece.

### Interior

Internally, at one level the building is a vast open empty space; a shed! At other times it is divisible into three sections each requiring individual lighting scenes. Recognising the diversity of uses and users, the challenge was to provide a valuable range of lighting modes that can be operated by a wide range of people with diverse technical skill, most of whom are likely to be unfamiliar with the facility. Lighting is applied in layers. The most basic layer is from functional CFL down lights for room set-up etc. mounted to custom lighting trays, rigged to the ceiling trusses.

The most significant layer of light for ambient scenes is delivered from high output cold cathode mounted in the custom lighting trays, that up light the ceiling. 2 colour temperatures are used; 2,700K and 3,500K, individually controllable to give a wide range of mood.

The walls provide the third architectural lighting scene. The logic used outside is applied inside the building, whereby the simple concrete columns are grazed with narrow beams of light from metal halide sources. Noting that the building is intended for live performance, infrastructure for theatre systems is also included comprising a matrix of lighting bars; distribution of power; DMX control ports; patchable outlets; programmable audiovisual control; stage & control connection points.

### Lighting control

Architectural lighting control

uses the Philips Dynalite technology.
There are 2 levels of user interface for architectural lighting.
For general, non expert use, custom panels are located in each division of the interior space. These have clearly marked descriptions; e.g. down lights; walls; ceiling etc….
For more sophisticated and site wide access, a touch screen is provided. This enables refinement of all the dimming options, overall control of the interior lighting as well as switching override for the exterior, which is otherwise time clock controlled.

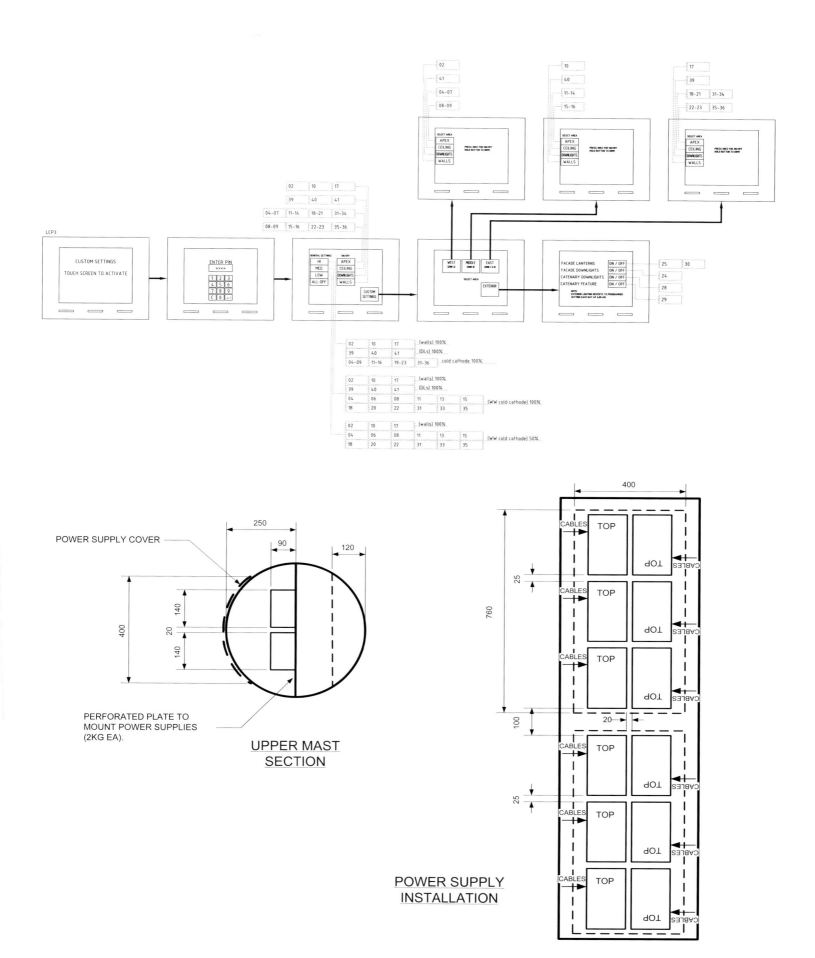

1号棚项目初衷是为霍巴特港务局建造一个仓库,项目的指导原则是将一个粗糙的功利性建筑转换为一个多功能公共大厅,以及向毗邻公主码头的极受欢迎的萨拉曼卡市区致敬。这个建筑约130米×35米,需要满足各种公共需求,如用于一年一度的"品味塔斯马尼亚岛"食物节、艺术活动、展览、训练和大型会议。前院需要一个大型的室外表演空间作为戏院和其他活动场所。建筑的目的是维持原有建筑的简单质朴,PointOfView将纯粹和基本的手法贯穿设计的始末。

照明设计的本质在于强调建筑作为满足各种功能要求的交流屏,也是夜晚当地环境的景观元素。通过突出常见的材料和建筑元素,一种强烈的空间感便在这座建筑的历史和环境中诞生了。

### 外部

1号棚主要是一个被看做对着开阔海域背景的大型建筑。港口的幽暗和相对别处较低的周围环境意味着非常少的光线就能制造出强烈、平衡的效果。立面设计由两个关键的照明设备组成,新甲板区域使用了柔和的光线,建筑的结构柱则使用了戏剧性的桅杆状灯具。

一条步行/自行车道被安装在天篷结构上的灯具照亮,对灯光效果深思熟虑后,这个发光体可能是建筑的全部亮点。

前院用来支持公共演出,有一个醒目的大型天篷,用4根超过16米的桅杆支撑。这种结构为公共区提供了照明(通过悬浮的LED罐),同时也是一种公共艺术;一块可编程的运动LED帆屏结合了声控系统,给当地的媒体艺术家们提供照明。艺术家是长期委任的,实际上他们有一个可编程的节点式RGB照明系统,以及可创造作品的音频媒体。

### 内部

建筑内部一层是一个巨大的开放空间,也就是一个棚。在其他时间它可以被分成3个部分,每个部分需要独立的照明系统。意识到用途和用户的多样性之后,难点在于给分散在各个角落的具有不同技能的人(大多数还不十分熟悉这个设施)设计出宝贵的照明模式。

照明设备被层层设置。基本层是功能性的下射灯等,安装了定制的灯盘,镶嵌在天花板结构上。

最值得一提的照明层是安装在定制灯盘里的高输出冷阴极灯,从上面照亮了天花板。分别使用了2 700K和3 500K的色温灯,独立控制,能提供广泛的基调。

在室外使用的逻辑被应用到建筑内部,简单的混凝土圆柱被金卤光源窄窄的光束扫过。

由于建筑是为现场表演做准备的,剧院系统设施一应俱全:照明棒的矩阵、能源配置、DMX控制端、补充插座、可编程视听控制、舞台和控制连接点。

### 照明控制

建筑照明控制采用Philips Dynalite技术。

建筑照明有2级用户界面。

对一般非专业用户来说,室内空间每个分区都有定制面板,并附有清晰的标识,如下照灯、壁灯、天花板等等。

为了更好地适应场地的宽度特意设置了一个触摸屏,汇集了所有的调光选项,整体调控室内照明及开关室外灯具,此外还有时钟控制。

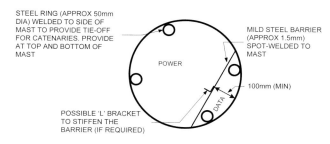

**MAST SECTION**

- STEEL RING (APPROX 50mm DIA) WELDED TO SIDE OF MAST TO PROVIDE TIE-OFF FOR CATENARIES. PROVIDE AT TOP AND BOTTOM OF MAST
- MILD STEEL BARRIER (APPROX 1.5mm) SPOT-WELDED TO MAST
- 100mm (MIN)
- POSSIBLE 'L' BRACKET TO STIFFEN THE BARRIER (IF REQUIRED)
- POWER
- DATA

**NOTES**
1. BARRIER PROVIDES MECHANICAL AND ELECTRICAL (NOISE) ISOLATION BETWEEN POWER AND SIGNAL CABLES.
2. MAST (AND BARRIER) EARTHED.
3. CABLES INSTALLED USING VERTICAL CATENARY

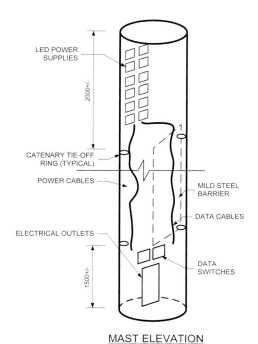

**MAST ELEVATION**

- LED POWER SUPPLIES
- 2000 +/-
- CATENARY TIE-OFF RING (TYPICAL)
- POWER CABLES
- MILD STEEL BARRIER
- DATA CABLES
- ELECTRICAL OUTLETS
- DATA SWITCHES
- 1500 +/-

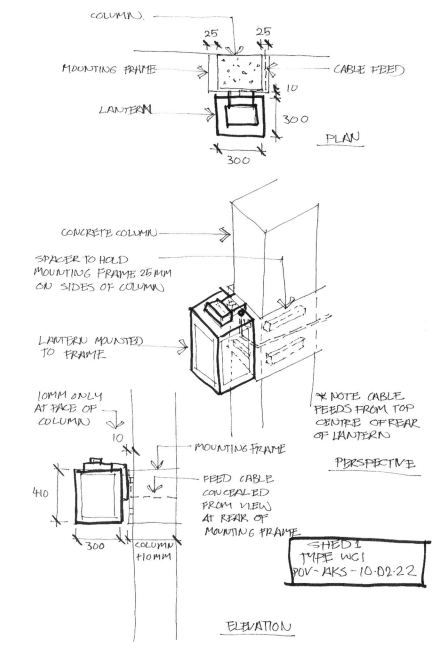

PLAN — COLUMN, 25, 25, MOUNTING FRAME, CABLE FEED, LANTERN, 10, 300, 300

PERSPECTIVE — CONCRETE COLUMN, SPACER TO HOLD MOUNTING FRAME 25MM ON SIDES OF COLUMN, LANTERN MOUNTED TO FRAME, *NOTE CABLE FEEDS FROM TOP CENTRE OF REAR OF LANTERN

ELEVATION — 10MM ONLY AT FACE OF COLUMN, 10, MOUNTING FRAME, 410, FEED CABLE CONCEALED FROM VIEW AT REAR OF MOUNTING FRAME, 300, COLUMN 110MM

SHED 1
TYPE WC1
POV - AKS - 10.02.22

# Louisville Second Street

落日的色彩
——路易斯维尔第二街
道交通和街景工程

**Credits**
Location: Louisville, USA
Owner: Downtown Development Corporation, Kentucky Department of Transportation
Lighting Designer and Color Palette Advisor: Leni Schwendinger Light Projects LTD
Landscape Architect: CARMAN

The road under the Second Street Bridge has been transformed into a plaza—filled with plantings, seats and pedestrian spaces to host festivals and celebrations—shaded by the dynamically illuminated overpass.

Leni Schwendinger Light Projects' streetscape scope included a service road and vacant land alongside the Clark Memorial Bridge. The historic cantilevered truss bridge, locally known as Second Street Bridge, crosses the Ohio River. Conveniently located adjacent to the new Yum Arena, the objective was to transform the off-ramp into a vibrant promenade. Schwendinger's goal for the illumination: "using simple, inexpensive—but innovative—means I wanted to transform the heavy infrastructure into a gently breathing enclosure for the new plaza space." Less than one-year after the contract was awarded, Light Projects' illumination and color design opened to celebrants in October 2010. A complex approval process was navigated, including federal

government agencies, state and city departments of transportation, and the local Waterfront Development Corporation, among others. The project was funded by the ARRA stimulus program, which called for a fast-tracked and economical design concept and solution.

Cross streets Washington and Witherspoon join Second Street lined with a row of wooden buildings on Washington which present their old timey "Whiskey Row" back doors to the street. The Iron Quarter buildings are being renovated into hotels, restaurants and bars. When Light Projects arrived, a generally disheveled, chipped and neglected sensibility pervaded the vicinity.

Light Projects participated in a design charrette in November 2009. The stakeholder workshop set the tone and direction for the design. "Bright" and "welcoming" were keywords for the lighting.

To create an inviting urban living room, Light Projects designed the illumination for seating areas, sidewalks and plaza. For the room's "ceiling", the underside of the iron bridge is enhanced with a floating effect of cast light, outlining and illuminating the I-beam surfaces and rivet textures. The duo-tone, red and gold color scheme is balanced with the cream color of paint coating. The colors—bridge as canvas and the lighting—celebrate Whiskey Row's heritage of amber liquid bourbon and colors of sunset.

Light Projects selected energy-saving fluorescent tubes for the bridge lighting—simple, industrial

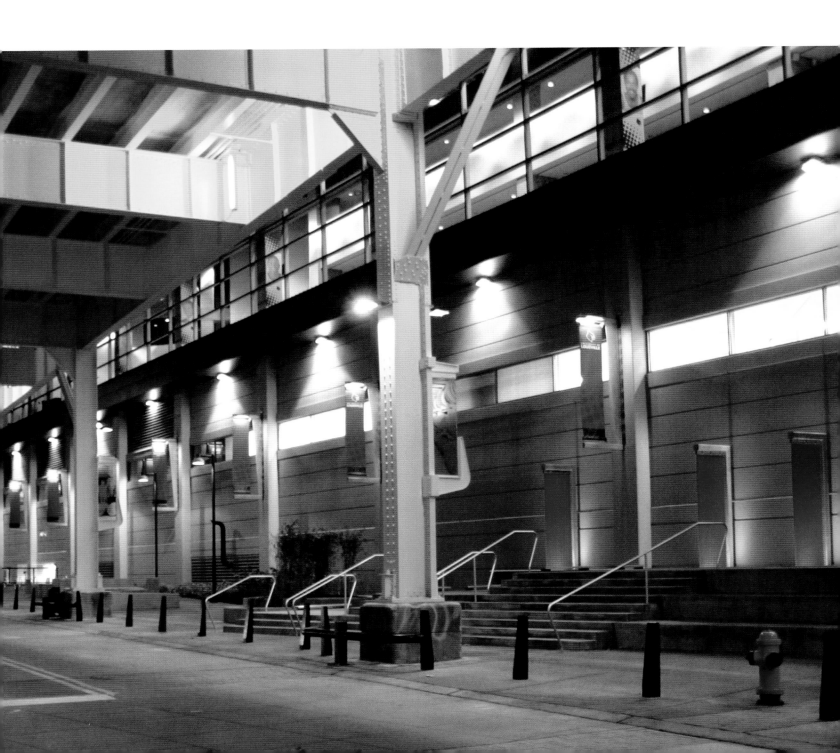

lighting fixtures with a twist —the luminaires were filtered with colored glass and controlled by Digital Addressable Lighting Interface (DALI). DALI is an electronic network protocol that is generally used for lighting in commercial buildings. Light Projects adapted it for exterior use and complex lighting sequences.

Another technological innovation is a line of bright beacons mounted on the face of the bridge, which are programmed in dynamic and rhythmic sequences each hour. Reconnecting with a French childhood friend turned lighting manufacturer, Schwendinger discovered LED flashers famed for lighting up the Eiffel Tower in Paris. The whimsical LED sequences countdown the hours, from sunset to midnight on weekdays, and sunset to 2:00 am on Friday and Saturday.

On the hour, nightly, the sequences bring the old iron structure to life; an invitation to visit a newly developed Whiskey Row, the plaza itself and Yum Arena.

第二街道大桥下的道路已经改造成一个广场，栽种了植物，设置了座椅区、步行区可供举办节日庆典活动。上方充满活力的照明高架桥为其提供了遮阴功能。

莱尼-史温丁格照明项目公司的街景设计范围包括一条服务道路和克拉克纪念大桥沿边的空地。这座具有历史意义的悬臂桁架桥横跨俄亥俄河，当地人昵称其为第二街道大桥。大桥毗邻新建的亚姆体育馆，设计目标在于将这个高速公路驶出匝道改造成一条生机勃勃的散步道。

史温丁格的照明设计目标：采用简单、廉价的手法，却不失创新，将重型基础设施改造成这个新广场空间的缓缓呼吸的围合物。

合同订立不到一年，照明项目公司就于2010年10月向参加庆祝的人们展示其照明和色彩设计。这一设计经历了复杂的审批程序，涉及联邦政府机构、国家和市级交通部门、当地滨水开发公司等。项目由ARRA刺激计划注资，该计划号召快速发展和经济型的设计理念及解决方案。

两条横向道路华盛顿街、威瑟斯彭街与第二街道交叉，华盛顿街有成排的木质建筑物，展现其昔日曾是第二街道的"威士忌街"后门。艾尔伦街区的建筑群正在翻新改建成酒店、餐厅和酒吧。在照明设计未启动时，这里的环境大体上是凌乱不堪的，周围也是备受忽略和遗弃的。

照明项目公司于2009年11月参加了一个设计研讨会。股东们提出了设计的基调和方向。"明亮"和"好客"是照明设计的关键词。

为了营造一个好客的"城市客厅",照明项目公司为座椅区、人行道和广场进行照明设计。"客厅"的"天花板",即铁桥底面,采用投光灯营造浮动的效果,勾勒并照亮L字形的大梁表面和铆钉纹理。红色和金色的双色配色方案,淡黄色的油漆层平衡。这些色彩,连同作为画布的大桥及其照明,是"威士忌街"遗赠的波旁琥珀酒和日落色彩的礼赞。

照明项目公司为大桥照明选择了节能的荧光灯管,简单的工业灯具略微弯曲,这些照明设备经过有色玻璃过滤,并由数字可寻址灯光接口(DALI)控制。DALI是一个电子网络协议,通常用于商务楼的照明。照明项目公司将DALI用于外立面和复杂照明组。

另一个技术创新体现在大桥正面的一排明亮的闪光指示灯,它们在每个小时都呈现动态而富于节奏的序列。史温丁格与法国的儿时伙伴重新获得了联系,这一伙伴如今成为灯具制造商,史温丁格得以发现照亮巴黎埃菲尔铁塔的著名LED闪光灯。这些怪诞的LED序列倒数着小时数,工作日是从日落时分至午夜,星期五和星期六则是从日落时分至凌晨2点。

夜间,在准点的时候,指示灯序列使旧的铁桥结构重获新生;邀请人们来此参观新发展的"威士忌街"、广场以及亚姆体育馆。

# Frederiksberg New Urban Spaces

铺展的画卷
——菲德烈堡新城市空间

**Credits**
Location: Solbjergvej, Frederiksberg, Copenhagen, Denmark
Client: Municipality of Frederiksberg
Lighting Design: SLA
Team: Stig L. Andersson, Hanne Bruun Møller, Stine Poulsen, Lars Nybye Sørensen
Collaborators: Hansen & Henneberg Lighting Engineers
Area: 18,000 m²
Cost of Construction: € 4,7M
Sponsors: The Realdania Foundation

With the desire to change leftover spaces between the bodies of prominent buildings into sensory spaces SLA has, with Frederiksberg's new urban spaces, created a surprising and vibrant urban area. The town centre consists of juxtaposed spaces with different characteristics. Appealing to the senses and with accessibility as a point of departure the five urban spaces contain qualities that we enjoy in nature: change, surprise and heightened awareness of transition and movement.

On Solbjerg Square luminescent mist floats close to the floor and circular reliefs cut into the surfacing collect rainwater into pools that reflect the sky. The soundscape is accentuated and varied by means of sound shafts placed around the square.

In autumn Falkoner Square is a profusion of red leaves covering the subtle change of level of the floor. With the Pinetum a bright, open park has been created with different kinds of pine on a grass surface. Rich in variation of scent and colour, and visually in the way the branches and needles move in the wind.

In spring Solbjerg Square vibrates with a shimmer of white flowers, where crab-apple trees and spruce stand side-by-side, their crowns lit up in the evenings: Light, scent, sound and movement. Together the five spaces and the passages between them form a whole.

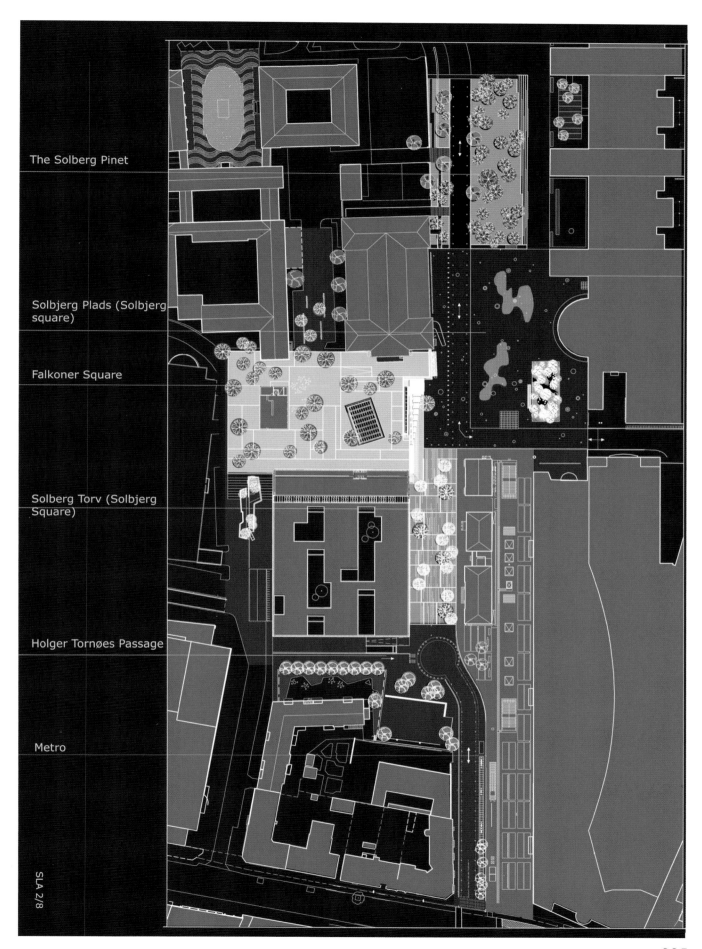

为了改造两座著名建筑间的围合空间，SLA设计事务所对此处进行了设计，使这里与周围城市环境相和谐，营造了奇妙而神秘的优美景观。这片中心区由一系列风格各异的空间共同组成。这五个城市空间具有我们从本质上享受的品质：变化、惊喜、对过渡和变迁的高度意识。

在索伯格广场上，发着冷光的雾气浮动在地面附近，雨水通过切入地面的圆形浮雕，汇入倒映出天空的水池中。广场周围各种各样的声音轴，丰富了人们的听觉享受。

入秋，飘落的红叶覆盖着变化微妙的地面。同松树园交相辉映的开放空间俨然成为了松树品种的展览馆。这里丰富的景致和色彩的变化给人以特别的空间享受，风中沙沙作响的针叶和树枝也吸引着游人的眼球。

在万物复苏的春天，索伯格广场上开出了白色的小花，掩映在山楂树和云杉之中。一行行的树木和花草在晚上摇曳生姿，光、影、声音和动静给人们源源不断的丰富感受。五个空间和它们中间的通路组成了这处完美的城市景观。

337

# 10th@Hoyt Apartments

点亮生活的夜晚
——霍伊特公寓

**Credits**
Location: Oregon 97209, Portland
Landscape Architect: Koch Landscape Architecture

In Portland, a city noted for pioneering and mandating use of new environmental technology, enhancing sustainability and mitigating environmental impact are critical goals. For this reason, Portland has already seen a proliferation of green roofs and LEED® certified buildings. This project creatively departs from typical details for stormwater collection, detention and quality enhancement through its Persian garden inspired sculpted forms, textures and colors. The design intent of the courtyard garden at 10th@Hoyt Apartments was to imaginatively display stormwater and provide extended detention of nearly all the roof top stormwater runoff.

The City of Portland requires developers to mitigate the impact of any increase in impermeable surfaces of developed property. Many engineered mechanical solutions have been developed to satisfy the guidelines, but without due consideration of the aesthetic appeal of water. This project explores the aesthetic possibilities and choreographs stormwater through a variety of playful and sculptural devices that continue to be a source of inspiration for landscape architects.

The design excites native Portlanders and building residents alike. The landscape and the integral stormwater features make users aware and appreciative of this natural resource by temporarily diverting stormwater from its typical path to the urban stormwater mitigation structures and

displaying, modulating and accentuating its delightful qualities. The entire landscape is built on a concrete structural slab over a parking garage limiting opportunities for infiltration and aquifer recharge within the project limits, but allows significant onsite detention. Stormwater can be detained for approximately 30 hours after each storm event, allowing time for sediments to settle out and the water to be cleansed naturally.

Three copper downspouts channel nearly all the roof stormwater into precast concrete runnels and chadars—two of which empty water into shallow detention basins, and a third emptying into a 40 foot long 4000 gallon detention cistern. Granite river rock adds textural contrast and secures the pools

from being a safety hazard. The basins and cisterns temporarily hold water and allow for recirculation over two sculptural Cor-ten weir boxes. Each water weir is also designed as a sculptural light feature that maintains interest wet or dry. Water in the cistern and basins begins to slowly weep into the district's stormwater system after the initial storm event. The total holding capacity of the basins and cistern allows the detention of roof water equivalent to a 1/8" rainstorm event. This allowed the downsizing of the conventional mechanical stormwater detention device used on the site.

The inner courtyard also provides semi-private respite for occupants and is open for public enjoyment. The central community seating area is cradled between the stormwater features and includes a section of covered seating so spectators can gather and watch during rain events. Evergreen, flowering and fragrant plants such as Camellia sasanqua 'Yuletide', Sarcococca rusifolia and Daphne odora take turns providing aromatic and color interest. Gunnera tinctora offers a surprise of exuberant texture and flower form for those who explore the courtyard. Stormwater that gathers from the encompassing building roof engages the entire courtyard space and presents a public display of light and energy. Even at night, the metal weirs pierced with glass buttons and illuminated by interior lights create a friendly space and add a lively presence to the otherwise out of sight - out of mind stormwater cycle. Night or day, rain or shine this courtyard remains a pleasant environment for residents.

在以环保而闻名的波特兰市,城市建设以提倡和采用新环境技术,提高城市建筑的可持续性,减少发展对环境的破坏为重要目标。因此,波特兰出现了越来越多的绿色屋顶和通过美国绿色建筑协会认证的环保建筑。霍伊特公寓的这项工程营造了一处有丰富的造型、结构和色彩的庭院花园。花园的设计目标是对所有从屋顶流下的降水进行引导和暂时的贮存。这项创造性的工程设计巧妙而细致,能够收集、贮存降水并对其水质进行净化处理。

波特兰市要求开发商尽量增加所开发房产的不透水外表。许多的解决方案达到了这一要求,但却忽视了项目尤其是水流的观赏性。而这座庭院花园的设计应用了多种能赋予景观设计师灵感的元素,通过对水流美学特点的研究,营造了一系列富于变化的美丽水景。

设计将建筑和波特兰当地特色融合到一起。精心营造的降水循环系统将降水从其路径引入城市降

水循环系统，并对降水进行调节和引导，突出了水令人愉悦的特点，让使用者喜爱上这处由景观和水循环系统组成的天然资源。所有这些景观都建造在停车场上方的混凝土地板上，由于受项目特点和位置所限，降水在这里只能短暂滞留而不能下渗。每次降水之后，降水可以在此停留30个小时左右，这段时间内，降水可以自然地沉淀和被净化。

三根铜制落水管把降水从屋顶引入预制的混凝土水道中，其中两根将水排入低浅的缓留池中，第三根将水排进容积达4 000加仑、长约12米的蓄水池中。池中的花岗岩石块不仅提供了视觉上的冲击，使其更加美观和迷人，同时也增强了水池的安全性。降水在这些水池短暂停留之后，从两个特色的水堰表面流过。不管是晴天还是水流充沛的雨天，精心设计的水堰都是一处迷人的灯光景观。降水之后，水池中的水都会缓缓地流入这个区域降水循环系统。这些水池的总蓄水量可以容纳一次约0.2米的降水的缓留。通过这些设计的实施，减少了区域内降水贮存设施的面积，增强了美感和整体的和谐性。

内部的庭院不仅为居民提供了半私人空间，还为公共休闲娱乐提供了方便。布局在降水循环系统中的社区座椅区安置了一些带顶篷的座椅，即使在雨天观赏者也可以在此相聚、闲谈，尽享生活的幸福与惬意。栽种的诸如茶梅、野樱桃、瑞香等常绿、开花、芳香植物使人们尽情感受鲜艳的色彩和扑鼻的芬芳。庭院中繁茂的植物和美丽的花朵也让游人体会到大自然带来的舒适和清新。从各座建筑屋顶收集来的降水为这处庭院提供了怡人的环境，展示了灯光和能量的优雅。夜晚，灯光从金属水堰的内部透射出来，灯光与水流交错，纹理优美的循环系统使这处庭院散发出别样的风情。不管是白天还是夜晚，晴天还是雨日，这里都是居民们钟情的休闲空间。

# The Albany Courtyard "Garden of Light"

## 光影欢乐园
## ——阿尔伯尼庭院

**Credits**

Location: Liverpool, UK
Landscape Architects: BCA Landscape

### Design Statement

Rescued from dereliction, the courtyard within this Grade 2* Listed Building in Liverpool in the UK has now become the setting for the ground breaking "Garden of Light".

The design team explored a variety of concepts, including habitable pods within the court, before eventually settling on a design based around a series of scrolling and spiralling contemporary forms. These were inspired by the Albany's original architect JK Colling and his passion for medieval foliage and flower illustrations, which can be seen in various motifs within the building's carved stonework and cast railings. The new seating, lights and trellis structures are all bespoke and unique to the Albany. In order to make the cutting edge design a reality and realise their vision, the designers enlisted the services of specialist furniture and lighting manufacturers from across Europe.

### Spiral Chandeliers

Every chandelier comprises 2,250 Swarovski Strass 14mm crystal beads set on a chromed helix. Each is approximately 1 metre tall by 1 metre in diameter, and weighs 25 kilos. Swarovski Strass crystal is considered to be amongst the finest quality crystal in the world, normally used for exquisite and luxurious jewellery. The chandeliers seem to float above the courtyard on a catenary wire system. To satisfy conservation concerns none of the stainless steel hanging wires could be above 4mm in diameter.

### Scroll Seating

Sinuous and sculptural seating constructed from moulded red GRP is situated directly beneath each chandelier at ground level. At night dramatic LED under-lighting makes the seats appear to hover above the original Yorkstone flags. The power for the lights and ground fixing details are all hidden within the spun aluminium legs. All the legs have varying diameters – scaling down as the section of the seat decreases. Each seat was constructed in 5 parts, so it could be carried through the front door and down a flight of steps in to the basement court.

### Trellis Cones

All the new elements within the court tread lightly on and around the original features. Rather than break out the courtyard paving for new tree-pits - the existing coal holes in the ground are utilised as planters - above which are fixed sixteen, 3 metre-tall trellis cones. As they grow, the ivies are being trained around the spiral. Each is illuminated with "shimmer"

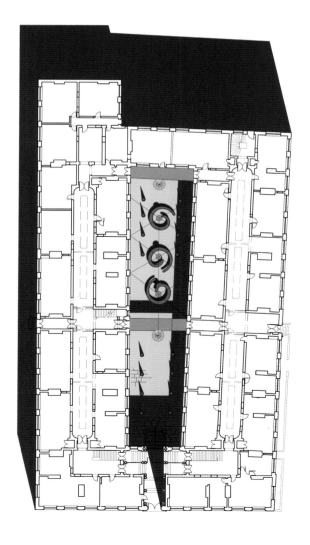

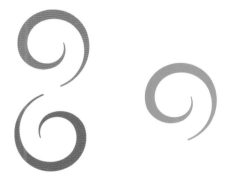

fibre optics which glimmer to accompany the seating and chandeliers. All the lights in the court can be controlled by a dimmer system within the concierge. Night-time viewing is essential.

### Restoration

On entering the courtyard, the visitor is met by the sight of the fully restored original cast-iron bridge that bisects the courtyard. The cast-iron spiral staircase that leads to the bridge is also illuminated with subtle up-lighting. The chandeliers and seating are revealed upon descending into the space - the new elements harmonising with the original features. The team worked closely with Heritage and Conservation experts within the city to ensure the careful restoration of original features and the retention of the view of the cast iron pedestrian bridge from the street. Within this context – elements such as the cast iron balustrading were refined to ensure compliance with modern safety regulations.

"We have always been very aware of the great heritage that we were working with here and hopefully we have shown that with care and attention to detail – new and old can exist together in harmony. Ultimately we wanted to add to the Albany's uniqueness and create something magical that celebrates the past, but looks to the future"—Andy Thomson (BCA Landscape).

## 设计理念

英国利物浦这栋二级历史建筑的庭院，经设计师的装点，一改往日破败的景象，成为"光影欢乐园"的背景地。

设计团队构思了许多理念，比如在庭院内放置可休憩的豆荚形状的物体，最终确定了设计以一系列的卷状和螺旋状的当代样式为基础。设计的灵感源自"阿尔伯尼"庭院原创建筑师JK Colling对展现中世纪绿叶和鲜花的热衷，体现在建筑物的石刻作品和铁铸栏杆上的各式图案中。庭院的新型坐凳、照明和格架结构都显得独一无二，是依"阿尔伯尼"庭院而预制的。为了使这一潮流尖端的设计得以实现，设计师在全欧洲范围内寻求了各种专业家具商和照明供应商的帮助。

## 螺旋枝形吊灯

每个枝形吊灯以一条镀铬螺旋线为骨架，自上而下垂吊着14毫米的施华洛世奇水晶珠，共2 250颗。每个吊灯的直径大约1米，高大约1米，重25公斤。施华洛世奇水晶是世界上质量最好的，多用于奢华和精美的珠宝镶嵌。枝形吊灯由吊索系统支撑着，仿佛漂浮于庭院之上。依建筑保护的要求，这些不锈钢悬链的直径都不超过4毫米。

## 卷状坐凳

正对着每个枝形吊灯，其下装置着螺旋盘绕的雕塑坐凳，由红色的玻璃纤维增强塑料制成。夜晚时分，坐凳下漫射出的LED灯光产生了鲜明的对比感，坐凳宛如盘旋在约克石铺的地板之上。灯光的电源和地面细部装饰都隐藏于铝制旋状凳脚里。凳脚的直径各不相同，根据坐凳面积的减小按比例递减。每个坐凳都由五部分组装而成，由此坐凳得以经过前门、螺旋楼梯而最后到达地下庭院进行组装。

## 圆锥形格架

庭院里所有的新元素都轻盈地落步在原初小品之上或周围。设计师并没有在庭院铺装上挖掘新树穴，而是利用地面上已有的煤洞作为树池，在其上装置了3米高的圆锥形格架。藤蔓植物将逐渐沿着格架螺旋地向上生长。光导纤维闪光灯散落在格架上，其散射出的光芒与坐凳和枝形吊灯遥相呼应。庭院内的照明由门房里的调光系统进行控制。夜景对观者而言是十分重要的。

## 修复工程

进入庭院，映入观者眼帘的是一座已修复的铁铸桥，这座桥将庭院一分为二，桥中央的一段铁铸螺旋楼梯连接着庭院，楼梯之上，桥板下方的微妙灯光照亮着这段楼梯。自楼梯而下，枝形吊灯和坐凳渐入视野，这处空间里，新元素与原初小品和谐地共融着。设计团队与这座城市的遗产保护专家密切合作，从而确保了对原初小品的严谨修复，以及街面铁铸行人桥风景的保留。在这一背景下，铁铸栏杆扶手等元素的精雕细琢，进一步保证了与现代安全条例的相符。

"我们总是提醒自己正在进行修复的是座伟大的遗产建筑，对细节的设计十分小心且全神贯注，极力展现新旧共融的风格。最终，我们为'阿尔伯尼'庭院的独特性格锦上添花，营造了某种魔幻的氛围，不仅赞颂历史，同时还展望未来。"——安迪•汤姆逊(BCA景观设计公司)

# "Vessel"

光之篮
——西雅图弗莱德·哈钦森癌症研究中心"容器"雕塑

"Vessel", completed in July 2008, is a centerpiece for the Fred Hutchinson Cancer Research Center. Rising more than four stories in a transparent and searching gesture, this monumental but delicate sculpture employs light to represent the optimistic spirit of the institution. It is a luminous container for the aspirations and hopes of all involved. By juxtaposing native plantings with crystalline structure, it suggests a variety of dualistic metaphors: natural and technological, intuitive and rational, transparent and opaque, formal and informal. This "basket of light" expresses the dynamism of the Center in the way light and shadows play off, through and around the combinations of materials. Its interwoven structure represents the interconnected and collaborative nature of the Hutchinson Center.

"Vessel" works on an urban scale, marking views along various axes, as well as on a human scale, allowing passage into and through the leafy core or

**Credits**
Location: Seattle, USA
Client: Fred Hutchinson Cancer Research Center
Dimensions: 39' diameter wide x 60' high
Lighting Design: C.E. Marquardt
Drawings & Renderings: Oanh Tran & Michael Gregg
Construction Drawings and Project Coordination: Oanh Tran
Project Administrator: Arleen Daugherty
Structural Engineering: Grant Davis and KPFF Consulting Engineers
Metal Fabrication and Erection: W.A. Botting Co.
Structural Rings: Albina Pipe
Cable and Fittings: West Coast Wire Rope
Glass Installation: Carpenter crew
Site Work: Lease Crutcher Lewis
Electrical: SASCO
Materials: Aluminum, stainless steel, laminated dichroic glass, beveled clear plate glass, concrete, landscape vegetation
Photography: Ed Carpenter

NORTHEAST SECTION / ELEVATION     EAST SECTION / ELEVATION

the sculpture itself. As the vegetation surrounding "Vessel" matures and honeysuckle vines grow up its lattice structure, the interior space will be extraordinary for its delicate light and combination of intimacy and monumentality. Hourly, daily, and seasonal changes in the light and vegetation will make the sculpture an abstract sundial as well as symbol of transformation. Its classic form, attractive materials, and hierarchy of scales will give "Vessel" universal appeal regardless of whether experienced from a passing car, adjacent building, or passage on foot through its center.

"Vessel" faced a challenging structural issue in that the site requires a tall sculpture to address axial views and to be in scale with surrounding buildings, but there are serious weight restrictions due to the load limits of the tunnel structure beneath. This dilemma was addressed with a design that is lightweight in spite of its monumentality. Employing aluminum,

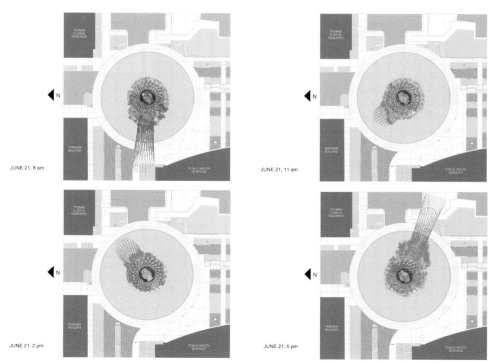

JUNE 21, 8 am     JUNE 21, 11 am

JUNE 21, 2 pm     JUNE 21, 5 pm

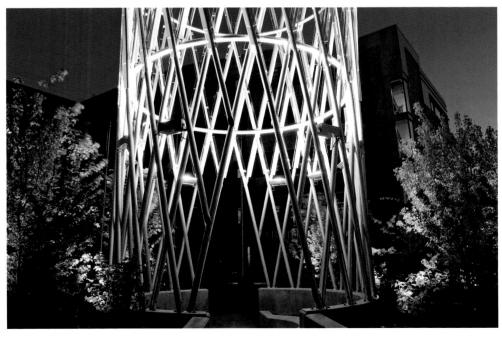

stainless steel, and slender strips of dichroic and beveled glass, the sculpture achieves both goals simultaneously. In an unusual innovation, laminated and tempered safety glasses were used structurally to strengthen the section of the aluminum members, allowing longer spans at lighter weight than with conventional methods.

"容器"于2008年7月完工，成为弗莱德·哈钦森癌症研究中心的中心装饰品。这个巨大却精致的透明雕塑，以追寻的姿态拔地而起，比四层楼还高，采用光元素象征研究中心积极乐观的精神。这个照明容器寓意所有与之有关的抱负和希望。将本土植物群和水晶结构并置，表达多样化的二元隐喻：自然的和科技的、直觉的和理性的、透明的和不透明的、正式的和非正式的。"光之篮"通过光影变化来表现研究中心的活力，光影穿梭在材料组合的结构间。相互交织的结构表征了哈钦森癌症研究中心内部相互联系、相互合作的本真状态。

"容器"为城市尺度的雕塑，沿着各种轴线框出景色，还包括人性尺度，即设置了进入和穿过叶状中心或雕塑本身的通道。"容器"周围的植物处于成熟期，忍冬藤爬上了格架结构，内部空间在精致的灯光照射下以及私密感和空旷感的结合中显得与众不同。时复一时、日复一日、季复一季，灯光和植物都在变化中，雕塑由此成为一个抽象的日晷，抑或转变的符号。雕塑经典的外形、引人注目的材料及其尺度向上的递增，都赋予它普遍的吸引力，无论是坐车经过、从邻近的楼房观望，或是徒步穿过它中心的通道。

"容器"在结构处理上遇到了富于挑战的难题，基地要求雕塑必须是高高的，这样能解决轴状面的问题，并与周围建筑物相称，而重量上却有严格的限制，因为基地下方的隧道结构的荷载十分局限。体量大、结构轻的设计，则解决了这个困境。通过采用铝、不锈钢、细长的磨斜边分色玻璃条，雕塑同时达到了前述目标。在一个独特创新里，安全度适中的复合玻璃在结构上增强了铝制构件部分，超出传统方法，使结构重量更轻，其上的跨度更长。

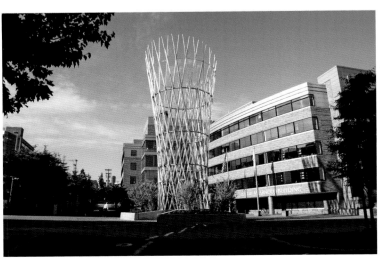

# City Lights Streetlight—Fabrication of Prototypes

城市之光
——城市街道照明设施雏型制作

**Credits**
Location: New York, USA
Client: The City of New York represented by New York City's Department of Design and Construction in partnership with the Department of Transportation
Lighting Designer: Office for Visual Interaction, Inc. (OVI), Jean M. Sundin, Enrique Peiniger
Structural Engineer: Werner Sobek New York - Werner Sobek, Maren Sottsmann
Manufacturer: Lighting Science Group Corporation
Electrical Engineer: Laszlo Bodak Engineers PC
Cost Estimator: Clare Randall Smith

### The City Lights Streetlight Project

In 2004, New York City's Department of Design and Construction, in partnership with the Department of Transportation, launched the City Lights Streetlight project. The international design competition to create a new standard streetlight for the City of New York drew over 200 entries from twenty-three countries, with multi-disciplinary teams including architects, engineers, urban planners, lighting designers, industrial designers and manufacturers. The new streetlight will provide a model for widespread lighting of streets, sidewalks and parks within the City's five boroughs. New York City intends to add the LED-based streetlight along with a high-pressure sodium version to the Department of Transportation's Street Lighting Catalogue, continuing a two-century long tradition of innovative street lighting.

The streetlight has already received an Arts Commission Award for Excellence in Design in 2004 and a Project Merit Award from the AIA, New York Chapter, in 2007.

### Designing with the Light Source of the Future

The new design will be the first addition to the New York City street lighting catalogue since the now-ubiquitous 250W high-pressure sodium Cobra Head was introduced almost fifty years ago. In creating a streetlight that will become a new classic, we asked ourselves, "What is the light source of the future?" Hi-flux LEDs emerged as an outstanding solution. With their small size, low wattage, intensity, and extremely long life of over 50,000 hours, LEDs are preeminent as an energy efficient, minimal-maintenance source. Standard in traffic lights and signage boards in 2007, high-performance LEDs have already appeared in luxury automobiles as daytime running lights, an indication that they have the brightness and stability to be used in streetlights. The streetlight is a vertical application with similar challenges to the horizontal illumination provided by headlights. Like the car headlight, the streetlight requires a long "throw" of light and a minimizing of glare.

After choosing high-flux, low-maintenance LEDs for the streetlight, we investigated a variety of configurations for the lights. Options including splaying LEDs outward, clustering the lamps, or arranging them in staggered lines to achieve the necessary light coverage. In the multi-lens model chosen, the LEDs themselves remain in a single, straight line, and the lens covering them is custom-molded to aim and focus the light as required. This configuration simplifies installation of the lights by eliminating the possibility of aiming errors, and provides the opportunity for an ultra-thin luminaire design.

The decision to use LEDs integrally shaped the form and aesthetics of the overall design. In contrast to the bulky luminaire heads associated with high-pressure sodium lamps, the LED streetlight takes on a slim, elongated profile enabled by the tiny size of its light sources, which do not require a hefty decorative enclosure. Instead, the thin arc of the luminaire itself provides the necessary surface area for housing and cooling the LEDs. The revolutionary design of the streetlight is specifically derived from the requirements and possibilities of LED technology. At the time of the competition, hi-flux LEDs were a completely new technology. Extensive calculations and laboratory measurements were undertaken to

demonstrate the new streetlight's ability to provide the light levels and distribution required by the city. Our calculations showed the LED streetlight's technical potential not only to match, but to outperform its predecessor. The LED streetlight provides a much more even and controlled distribution of light, free of lighting 'hot-spots' and neatly directed along the length of the street.

**Modularity and Technological Advance**

The streetlight uses LED modules, each installed behind primary and secondary optical systems. These work in tandem to focus and direct the light, achieving the required distribution pattern and illumination levels to meet city requirements. This design strategy allows the same LED segments to be used in a variety of streetlight configurations. The interchangeability of LED components within each fixture facilitates fabrication, installation, and

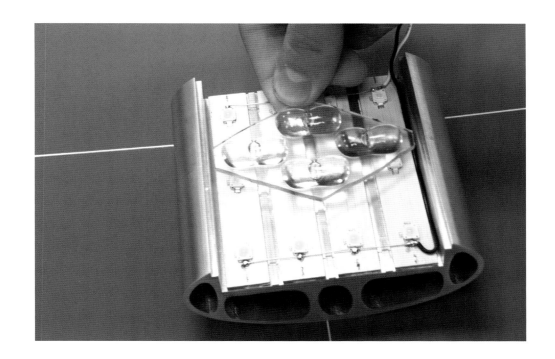

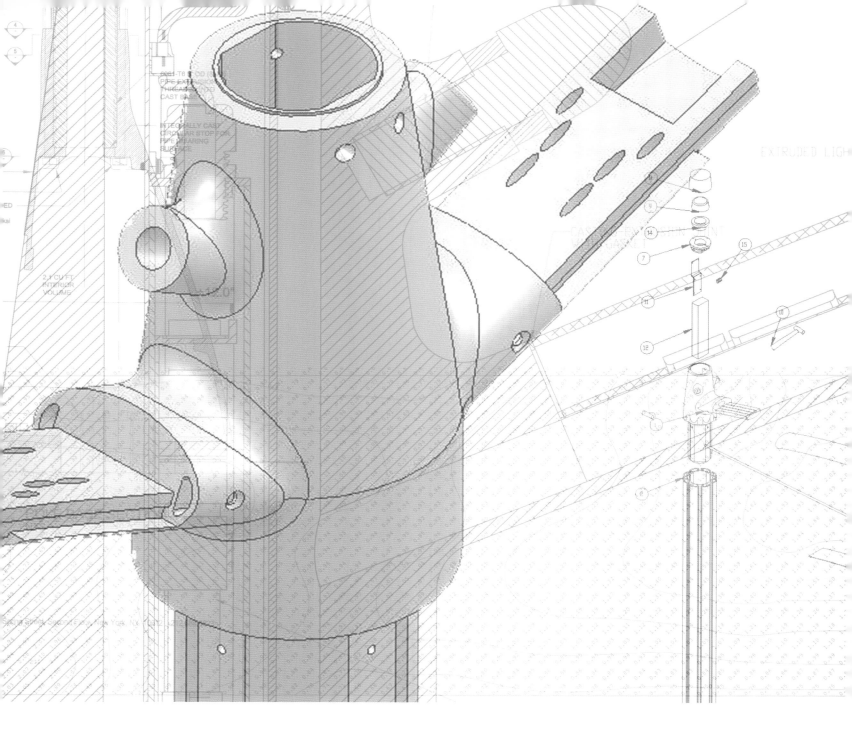

maintenance.

The use of a modular system also builds future flexibility into the system. LED technology continues to improve at a rapid rate, and the streetlight is inherently designed to accommodate this change. The streetlight has the ability to advance with time, becoming less costly and more energy-efficient as technology evolves.

### Rethinking the Streetlight: From the Ground Up

Allowing a one-to-one replacement with existing equipment, the streetlight is anchored to foundation bolts already cast in city sidewalks, and uses the central conduit connection from the former installation.

Specific attention was given to the method of fixing signage, call boxes, and other components to the

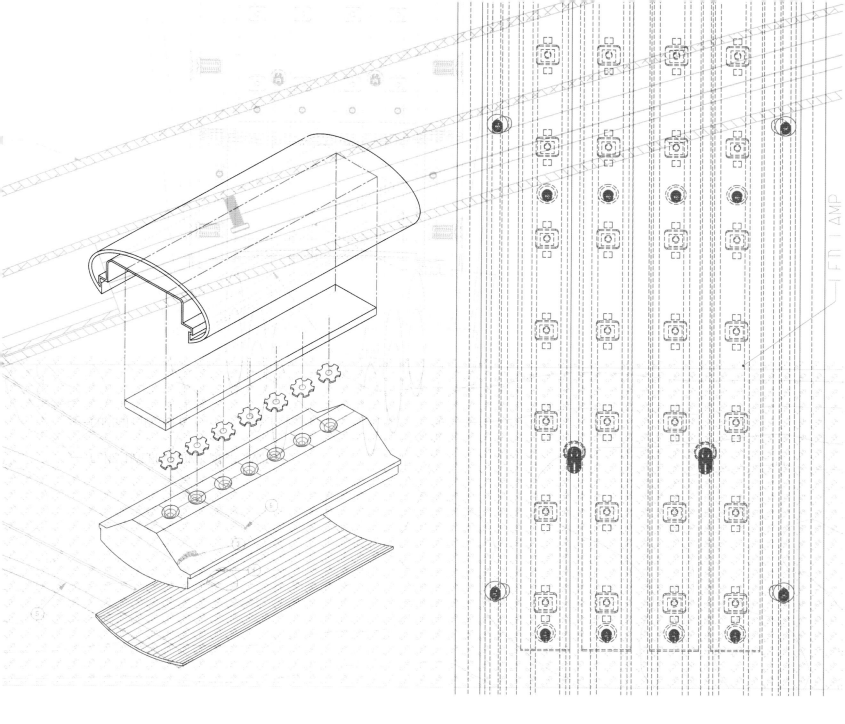

support pole. The pole features a fluted profile, created to work in conjunction with a new mounting system for signage. Components can be slid or snapped into place at any location along the length of the pole, at any orientation. This provides a clean visual appearance, in contrast to the metal bands currently used to strap signage onto poles. The fluted design reduces opportunities for vandalism by minimizing flat surface areas.

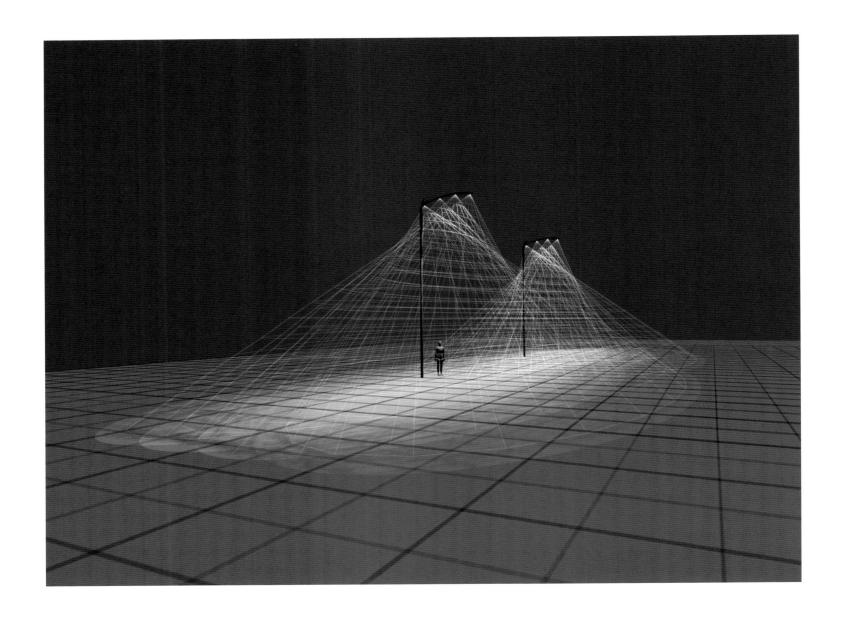

### 项目概述

2004年，纽约市设计施工与交通运输部门联合启动城市之光街道照明项目，其目标是为纽约市区建立一个全新的标准化街道照明系统。入围项目竞标的设计作品多达200余个，分别来自23个国家的各类团体组织，其中包括建筑设计公司、工程师团队、城市规划组织、照明设计企业、工业设计团体以及相关的制造商。

全新的街道照明系统将为纽约市内五个城区的街道、人行道及公园等公共场所提供广泛的照明设施，该系统将成为同类项目的典范。纽约市政府计划在交通运输部街道照明目录中所列的高压钠街道照明标准的基础上，新增LED照明系统，以继承和发扬街道照明创新长达两个世纪的悠久历史传统。

该项目曾荣获综合艺术委员会2004年授予的卓越设计奖及2007年由美国建筑师协会纽约分会颁发的优秀设计奖。

### 设计理念

该项目将成为纽约街道照明目录建立以来的第一个新增的设计类别，目前普遍应用的250瓦高压钠灯Cobra Head还是50年前投入使用的。当设计一种创新性街道照明设施，并希望它能够成为同类设计中新的经典之作时，设计师们自问："未来的光源应该是什么？"高通量LED作为一个出色的解决方案出现了。由于其体积小、功率低、强度高，加之高效节能、低维修率以及超过50 000小时的超长使用寿命等优点，LED无疑是最为理想的街道照明设施。

在2007年出台的交通信号灯与标识牌标准中，高性能LED已经广泛应用于豪华汽车的日间行驶照明中。有迹象表明它们在街道照明使用中的亮度和稳定性良好。街道照明是对照明设施的一种垂直应用，但它却与汽车前灯的水平照明存在着类似的问题。如同汽车前灯，路灯同样需要较长的投射光线与最大限度减少眩光的性能。

在选定高通量、低维护的LED材料作为本项目的主要材料后，设计师们对其各种不同的排列法做了详尽的分析：将其在户外分散放置、集中放置，亦或是交错放置，以满足必要的光线覆盖面。在多透镜模式中，LED本身就处于一种单一的直线照射状态，而覆盖其上的透镜都是定制模塑而成的，以汇聚充足的亮光。这种配置通过消除光线对准误差的可能性，简化了照明设施的安装，并为超薄泛光灯具的设计创造了机遇。

LED材料的使用形成了整体设计的完美形式和视觉美感。相较于高压钠汽灯相对笨重的泛光灯体，LED路灯虽体形修长，但其光源充足，且无需厚重的外部灯罩作为装饰。反之，轻薄的弧型灯罩可对LED灯体表面起到有效的保护和冷却作用。无疑，街道照明的革命性设计源于对LED技术的迫切需求及其自身的发展潜力。

在项目的竞标过程中，高通量LED是一种全新的技术。大量的计算与实验室测量结果都证明了新型街道照明设施的高照明度足以供给城市公共区域的照明需求。设计师们的计算显示，LED街道照明设施的技术潜力远远超过了传统照明设施，它的光线更均匀，光束分布面积更广，且不具照明"热点"，使得街道更增整洁之感。

技术发展

街道照明设施使用的LED组件都安装在一级与二级光缆系统的后面，它们进行协同运作，有效地控制了光的集中度和方向，实现了街道照明的合理分布，提升了照明水平，满足了城市发展的需求。

这种设计策略使得各种路灯配置都可以使用相同的LED材料。每个灯具内的LED部件可灵活互换，便于制造、安装和维护。

这样模块化系统的应用也为系统本身的维护创造了灵活性。随着LED技术不断地快速提高，街道照明设计也将顺势更加完善，从而设计出更为低成本、低耗能的城市照明设施。

观念转型

新型街道照明设施可与传统设施进行一对一替换，且无需移除已镶铸于城市人行道上的用于固定照明设施的地脚螺栓，还可延用旧有照明设施的中枢套管接线。

特别值得一提的是在路灯支撑杆上安装固定标牌、公用电话箱和其他组件的相关方法。支撑杆上设有一个凹槽配置，是用于连接安装固定标牌等组件所使用的新装置。这些组件可在支撑杆上的任何位置以任何方向灵活地安装到位。这样的设计方法为街道提供了简洁的视觉外观，显然比目前采取的金属捆带将标牌等组件捆绑固定于支撑杆上的方法更为科学和美观。此外，凹槽设计还通过最大限度地缩小了平坦表面的面积，减小了乱涂乱写对路灯柱的破坏。

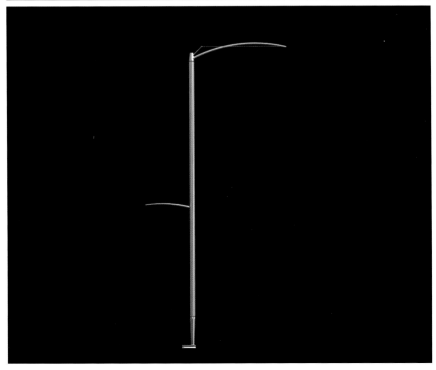

# "Tecotosh" Portland State University

## 结构与光线的互动
## ——波特兰州立大学"Tecotosh"雕塑

This sculpture is a graphic illustration of the combination of four basic engineering principles: tension, compression, torsion, and shear. Its title, "Tecotosh", is composed of the first two letters of each of those terms, and its structure is intentionally provocative from an engineering standpoint. The sculpture's structure has been conceived and developed in collaboration with structural engineers Grant Davis and Bob Grummel.

"Tecotosh" creates a gateway to Portland State University and the Maseeh College of Engineering, and provides an enormous gesture toward Lovejoy Fountain. Like an arbor, it creates a space beneath for sitting. It is highly visible from 4th Avenue, and from inside both adjoining buildings. Lighting is integral, creating a dramatic presence at night. The big curving form of the sculpture makes graceful counterpoint to the surrounding architecture. "Tecotosh" is animated by the incorporation of laminated dichroic glass playing in the light. Sunlight is projected and reflected from the glass in bright kinetic patterns cast across the landscape and architecture. Daily and seasonal movements of the sun passing through the sculpture create color patterns and an enormous shadow like a very abstract sundial.

This interaction of structure and light represents the joining of the engineered world with the natural world; science with nature. We understand the behavior of materials, structures, and light; but when they are combined in this way there remains an overriding sense of mystery-appropriate to the fact that engineering at its best is both a science and an art.

**Credits**
Location: Oregon,USA
Client: Oregon University System
Lighting Design: Craig Marquardt
Project Manager and Working Drawings: Oanh Tran
Project Administrator: Arleen Daugherty
Architect: ZGF, Portland, OR
Maseeh College of Engineering and Computer Science
Portland State University
Landscape Architect: Walker & Macy
Engineering: Bob Grummel and Grant Davis, Portland, OR
Metal Fabrication: Albina Pipe Bending, Portland, OR
Glass Fabrication: Haefker Studio, Portland, OR
Materials: stainless steel truss, laminated dichroic glass, stainless steel cables and hardware, aluminum light housings, up and down lights.
Photography: Ed Carpenter, Bruce Forster

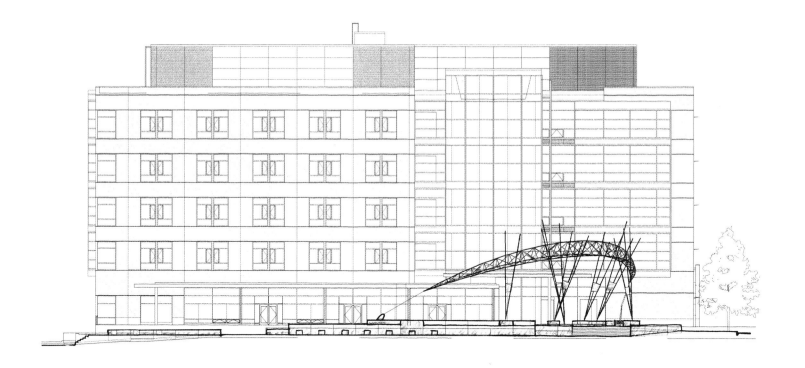

这座雕塑是四个基础工程原理相结合的生动演示：张力、压力、扭力、切变力。它的名称由这四个词语的头两个字母组合而成。其结构意在激发人们对工程方面的思考。雕塑结构的构想和设计得益于和结构工程师格朗特·戴维斯、鲍伯·格鲁米尔的合作。

"Tecotosh"为波特兰州立大学麦西哈工程和计算机科学学院创造了一个入口，在爱悦喷泉附近竖立了一个巨大的物体。雕塑宛若凉亭，其下的空间为人们提供了座椅区。从第四大道可以清晰地看到这座雕塑，从校园内附近的教学楼也可看到它。照明成为雕塑不可分割的一部分，在夜晚创造了富于张力的效果。雕塑大大的弧线外形与附近的建筑物遥相呼应，十分优美。

"Tecotosh"整合了夹层双色玻璃，光影中极富动态感。阳光投下并被玻璃反射，从明亮的动力图案延伸至景观和建筑。日复一日，季复一季，随着太阳的移动，光线穿过雕塑，创造出彩色图案，其投射的巨型影子仿佛一个超级抽象的日晷。

结构和光线的互动，表征着工程设计领域和自然界之间的联系；科学和自然的联系。我们了解材料、结构和光线的性能，但当它们以此种方式组合起来，仍然存在着压倒一切的神秘感，与这样一个事实相合：工程设计只有在整合了科学和艺术两种元素后才算是完美的。

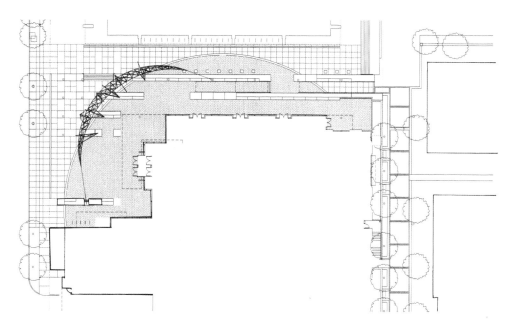

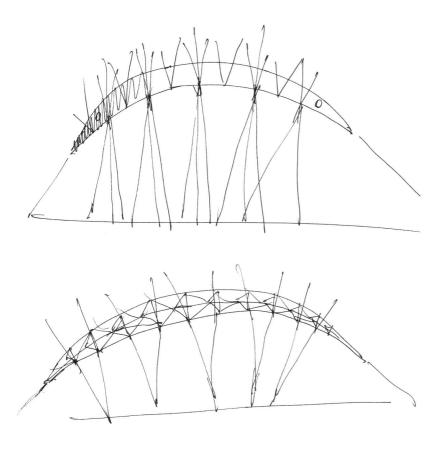

# Lauderdale-Hollywood Airport, Florida

"鳍扇"
——劳德代尔堡-好莱坞国际机场雕塑

**Credits**
Location: Florida, USA
Client: Broward County
Dimensions: 80' x 200' x 50'
Lighting Design: Craig Marquardt
Materials: tapered aluminum masts with white powder-coated finish; custom fabricated adjustable stainless steel anchors; laminated dichroic safety glass fins; landscape materials; lighting
Drawing & Renderings: Oanh Tran & Robert Lochner
Project Administrator: Arleen Daugherty
Structural Engineering: KPFF Consulting Engineers, Portland, OR

Rising into view on approach to Fort Lauderdale-Hollywood Airport, a gleaming structure presents itself on the horizon. Its eighty-foot spines gesture toward the runways. Like an enormous wing or fin or fan, Finfan presents an iconic form, suggesting images aquatic, botanical, mechanical, aerodynamic. Its towering scale insures prominence against massive infrastructure; its transparency gives delicacy in a ponderous landscape. From the air and roadways, Finfan is guardian and gateway. At night it glows mysteriously, hues slowly changing, illumination sliding across its outstretched ribs. Uncompleted due to site permit issues.

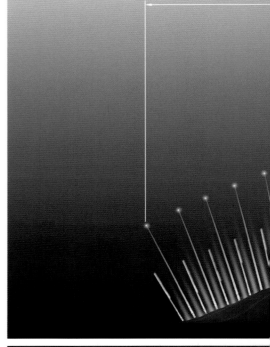

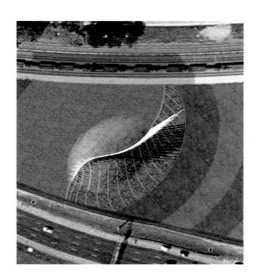

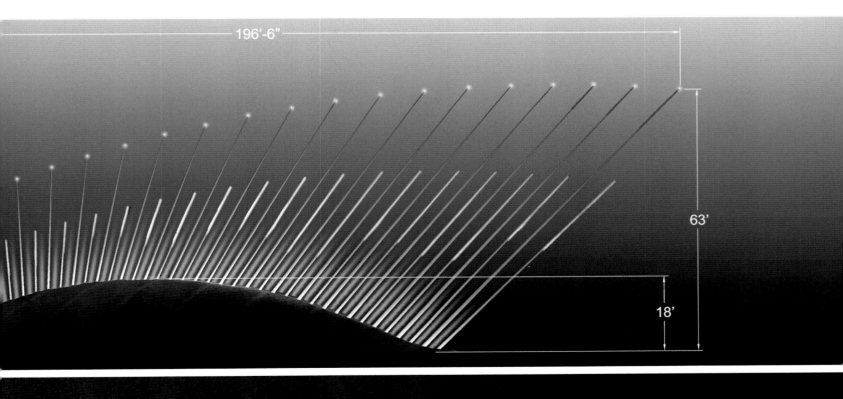

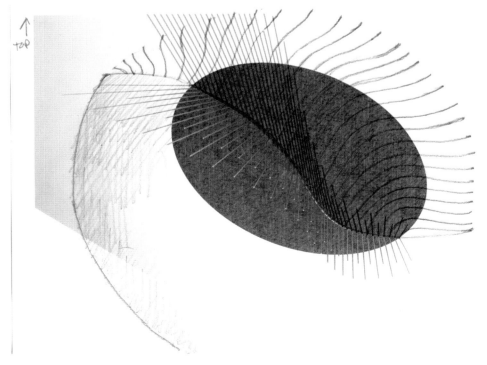

在通往劳德代尔堡-好莱坞国际机场的入口处，一个发光的结构在地平线上拔地而起，进入人们的眼帘。约24米长的脊柱指向停车场车道。"鳍扇"仿佛巨型翅膀或鱼鳍或扇子，成为一个标志性结构，表征了水生的、植物的、机械的、空气动力学的形象。它高耸的规模确保了它面对巨大的基础设施也毫不逊色；它的透明性为周围沉重的景观注入优雅元素。从空中到地面道路，"鳍扇"是守护者和门户。夜晚，雕塑发出神秘的光，色调缓慢变化，灯光滑动于整个伸展的翼肋。由于基地许可问题，项目尚未完工。

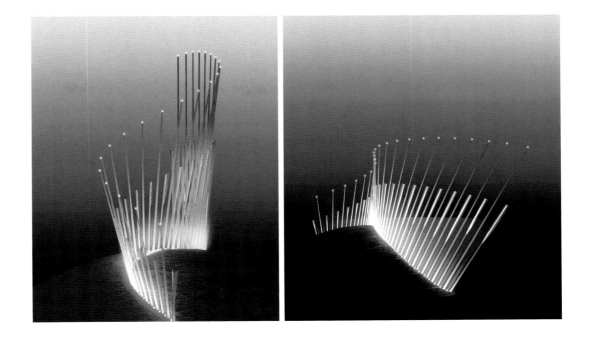

# "Triplet"

## 悬浮的三角研究园
## ——罗利达勒姆国际机场三联雕塑

**Credits**
Location: North Carolina, USA
Lighting Design: Craig Marquardt
Dimensions: Masts are 36', 43', and 48' long, 9" diameter at center, tapering to 4".
Materials: tapered hollow laminated wood masts, welded aluminum fittings, laminated glass tip elements, dichroic glass in anodized aluminum extrusions, stainless steel cables and hardware, LED lighting
Project Assistant:
Renderings, CAD drawings: Oanh Tran
Project Administrator: Arleen Daugherty
Structural Engineering: Grant Davis; Robert Macia, Stewart Engineering
Installation: Ed Carpenter, John Rogers, HannsHaefker, Craig Marquardt, Grant Davis
Cables and Fittings: West Coast Wire Rope
Wood Masts: Hennessey House, Sierra City, California
Glass Fabrication: HannsHaefker
Electrical Installation: TruePower Electric
Art Consultant: Wendy Feuer
Architect: Fentress
Photography: Ed Carpenter

"Triplet" creates a powerful presence in Terminal 2 at Raleigh Durham International Airport, positioned to be seen from long distances on either axis of the ticketing and baggage levels. Revealing increasing complexity as one approaches, it forms an engaging canopy that is experienced freshly from every angle whether moving up or down through the Central Opening. "Triplet" is suspended only from six anchor points integrated into the second floor railing stanchions, leaving viewers with a sense of awe at the apparent weightlessness of the installation. Triangular forms created in space suggest the Research Triangle, and refined, hand-finished materials and fabrications are reminders of North Carolina's legacy of fine craft.

The sculpture's immense Douglas fir masts are crafted from the same material as the building's impressive beams overhead. Dichroic glass elements provide rich color and slowly moving light projections when sunlight strikes through clerestory windows. Mast tips and anchors glow with embedded L.E.D. lighting. The floor beneath the sculpture is designed with a complimentary terrazzo pattern emphasizing the points at which art and architecture merge.

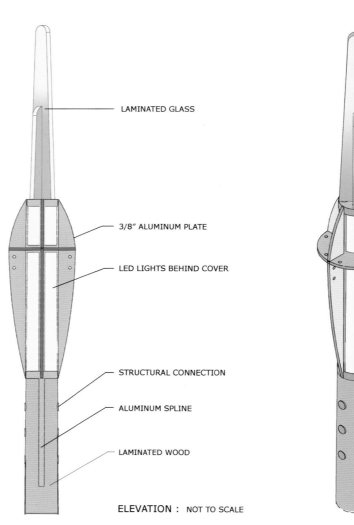

ELEVATION : NOT TO SCALE

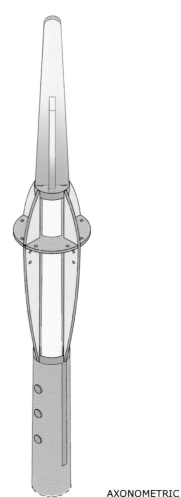

AXONOMETRIC

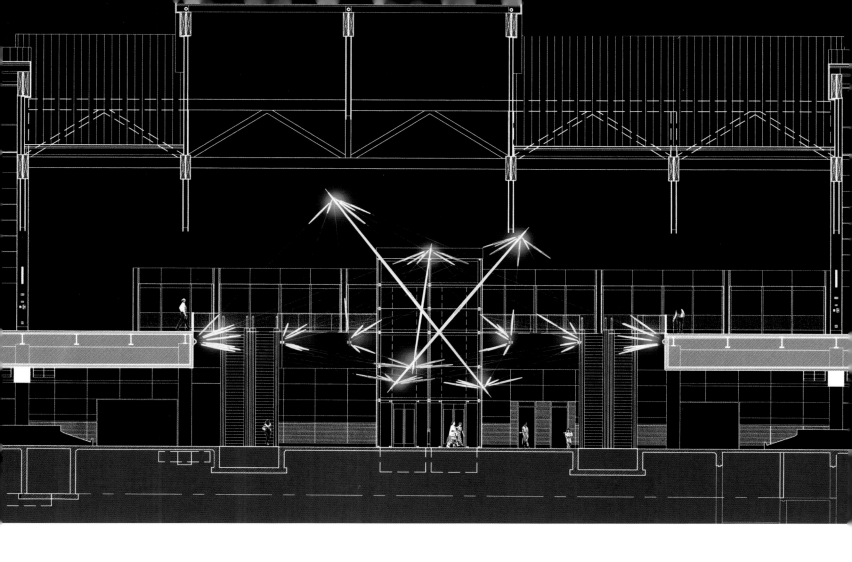

"三联"雕塑装置在罗利-达勒姆国际机场的2号航站楼,极具视觉冲击力,即使从检票层和行李提取层的轴线处仍能望到这一装置。越靠近这一雕塑,越能看到其所展示的复杂性。雕塑形成一个美丽动人的天蓬,当它在中央通道上下移动时,从不同角度都能体会到别样的新鲜感。"三联"悬浮于空中,仅用六条锚定位于二层栏杆支柱,以此令观看者对于装置外观上的无重状态心生敬畏。在空间里创造出的三角形寓意"三角研究园",经过改良的、手工修整的材料和构造物,提醒着人们关于北卡罗来纳州的精致工艺遗产的回忆。雕塑的巨型黄杉木柱杆,用与航站楼顶部横梁相同的材料加工而成。二向色玻璃元素在日光透过天窗照射进航站楼的时候,展现了丰富的色彩和缓慢移动的光投影。柱杆尖部和锚定部分运用嵌入式LED灯点亮。雕塑下方的地面采用水磨石铺装图案,与雕塑相呼应,突显了艺术和建筑的共融点。

# Gifu Kitagata Apartments

光、色、艺术、生活
——日本岐阜kitagata公寓

**Credits**
Location: Gifu-ken, Japan
Landscape Architects: Martha Schwartz, Inc.

This courtyard project is part of an experiment in "feminism in housing design" which also includes four apartment buildings designed by Akiko Takahashi, Kazuyo Sejima, Christine Hawley, and Elizabeth Diller. In the project master plan, the courtyard lies between the four separate housing blocks designed by these architects. Because of the diversity of architectural design found within the project, strong site imagery and geometry have been created for the courtyard to unify the distinct parts of the project and to give the project a memorable identity.

Before its present use for housing, rice paddies existed on this site. The geometric pattern of raised dikes and sunken paddies provides the metaphor for creating a series of sunken garden "rooms". These rooms offer a variety of opportunities for passive enjoyment or active play including water features,

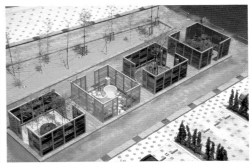

children's play opportunities, and public art. In the Willow Court, a sunken, flooded area with willow trees and wetland vegetation is made accessible by a wooden boardwalk. The Four Seasons Garden is a series of four miniature gardens that capture the spirit of each of the seasons and are enclosed by colored glass walls. In the Stone Garden, a circular fountain with stepping stones and rocks that spit water at irregular intervals creates a children's play pool. The other garden rooms are the Cherry Forecourt, Iris Canal, Dance Floor, Children's Playground, Sports Court, Water Rill, and Bamboo Garden. Each of these rooms provides a different experiential opportunity for the people who live in this community.

这个别致的庭院项目属于"女性主义住宅设计"中一个实验的部分，其他还包括由高桥晶子、妹岛和世等、克莉丝汀·霍利以及伊丽莎白·迪勒设计的四栋公寓大楼。在项目总体规划中，庭院位于这四个分开的住宅区中间。鉴于各个部分的不同设计风格，为了在统一之中赋予该庭园别具特色的吸引力，景观设计师们在设计中融入了几何特色以及强烈的意象，让这片静态的土地焕发出活力和光彩。

在此地成为住宅区以前，曾经稻浪滚滚，那些浮起的堤埂和凹陷的稻田为设计提供了形象化的设计素材，在此基础上创造出了一系列下沉式的花园小空间，为人们的休闲娱乐提供了多种可能。人们在这里可以尽情欣赏水景，饱览公众艺术，而孩子们则在这里找到属于自己的乐园。柳院内有一块凹地，里面种植着柳树和湿地植被，由一条木板路通往。四季园由四个小型花园系列组成，体现了四季之景，被由彩色玻璃组合而成的开放式景观墙包揽在内，无论是色彩还是造型，都颇具魅力。在石园内，踏脚石和大石块被布置在造型独特的圆形的喷泉中，水珠时不时地跳跃起来，异趣横生，孩子们在这里嬉戏玩耍，乐趣陶陶。此外，人们还可以遍览樱园、鸢尾园、游戏场、运动场、小溪以及竹园。丰富的景致让人大饱眼福。

# The Port Pavilion

## 光影潮汐
——圣地亚哥百老汇港口大厅

**Credits**
Location: San Diego, California
Dimensions: 45'h×16'w (wall sculpture)
Artist: Leni Schwendinger
Design Studio: Leni Schwendinger Light Projects, LTD
Fabricator/Installer: Basile Studio
Custom Cast Glass: Architectural Glass Art
Materials: Architectural Glass Art, Inc., Forms and Surfaces, Electronic Theatre Controls, Martin Architectural, Altman Lighting
Medium: textured stainless steel, custom cast glass, light projections
Project Cost: $418,000

The Port Pavilion, at the tip of the Broadway Pier in San Diego, opened in December 2010. Light Projects designed the illumination for the San Diego Bay North Embarcadero as part of the collaborative design team for this waterfront esplanade. Following that job, the Port of San Diego's public art division awarded Leni Schwendinger a commission for their minimalist glass-and-metal cruise terminal building. A 45-by-16-foot wall sculpture, Tidal Radiance, was designed in concert with an environment of projections to evoke tides and sea life. Featured on the building's exterior, custom cast glass and metal forms are mounted to a unique framework of tension wires.

To set the stage for Tidal Radiance's dramatic and luminous transformation during the darkened hours, it was important that its sculptural materials remain neutral by day. To accomplish this, Light Projects collaborated with the architect on material selections and other media that influence light and shadow qualities by sunlight.

After dark, Tidal Radiance becomes a shimmering, organic form with artwork elements seen from near and far. The artwork is visible to pier visitors, boats,

pedestrians and motorists along the Embarcadero promenade.

The interplay between viewer and artwork is integral to Tidal Radiance. The spectator becomes a participant through spatial and sensory immersion in the luminous stenciled projections along the entry and perimeter of the terminal. This environment is composed of a complex line drawing, which is echoed in the cast glass of the wall sculpture. Through the medium of light, chromatic seasonal compositions evocative of San Diego Bay culture and environment materialize. Luminous greens express the whale watching season, and glowing golds the cruise season. During the moon cycles, the full moon is expressed and emanates pale blues, while the new and quarter emanate deep and medium blue hues. Like all of Leni Schwendinger's work, Tidal Radiance contains the element of change. Whether animated patterns or a calendar of seasonal light sequences, one of the continuing challenges is to utilize the property of light to brighten, fade, and disappear – and to respond to controlled voltages through highly sophisticated computer programming. This element of controlled changeability – combined with color symbolism – allows the artist to create public art that not only pleases the eye but communicates and displays nuanced messages about the environment we live in.

圣地亚哥百老汇码头的港口大厅，于2010年12月对外开放。莱尼-史温丁格照明项目公司作为滨水散步道的设计团队之一，负责圣地亚哥湾北部散步道的照明设计。其后，圣地亚哥港公共艺术部门委托莱尼·史温丁格为其极简主义风格的玻璃金属轮渡大厅进行照明设计。

45英尺×16英尺（约13.7米×4.9米）的"光影潮汐"外立面雕塑，通过投影于所处环境，唤醒人们对潮汐和海洋生活的回忆。为了突显大楼外立面，特别制作了压铸玻璃和金属物件，将其挂在一个独特的高强度铁线框上。

为了使"光影潮汐"在夜间大放异彩且富于光影变化，确保雕塑材料在日间保持素颜显得十分重要。由此，照明项目公司和建筑师合作，精心挑选材料和其他会受日光照射而影响光影效果的介质。

夜幕降临，由远及近、由近及远，都可以看到"光影潮汐"闪烁的有机形象。这件艺术品的可见范围覆盖了滨水散步道沿途的码头观光客、船只、行人以及驾车的人。

观看者和艺术品之间的互动成为"光影潮汐"不可分割的一部分。观众通过空间和感觉的介入，成为轮渡码头入口和周围空间照明投影的参与者。一系列复杂的线条画构成周围环境，与外立面雕塑的压铸玻璃交相辉映。

通过光这一介质，实全化的彩色季相元素唤醒了人们对圣地亚哥湾的文化和环境的记忆。发光的绿色代表观鲸季，鲜艳的金色代表邮轮季。月亮周期变化里，满月散发出淡蓝色的光，新月和弦月散发出蓝色至深蓝色的光。

"光影潮汐"和莱尼-史温丁格照明项目公司设计的其他作品一样，整合了变化的元素。无论是动态图案或季相光指令序列历法，设计所面临的持续挑战之一是：利用光的性质实现变亮、变暗和消失的效果，同时通过高度复杂的计算机程序响应被控电压。控制中的可变性元素和色彩象征相结合，使艺术家创造出的公共艺术品，不仅悦目，还能传递和展现人们生活环境里微小的信息。

# "De Beemd" in Velp

光束波浪舞
——威尔普德比姆工业区

**Credits**
Location: industry area "De Beemd" in Velp
Client: city of Rheden
Project Artist: Ilona Lénárd
Design Team: Ilona Lénárd, Kas Oosterhuis, Chris Kievid, Cas Aalbers

Goal
Creating a real-time artwork for office-area.
Inspiration
The work is inspired by the movement of the continuous waves on the water surface and gracious swing in the wind.
What is it?
It's a series of flexible hollow tubes which more than 4 meters. The diameters are narrow enough to ensure a big waving.

How does it work?
The top of the stalk can be turned on or off by computer, and the light intensity is adjustable.
Place
The series of swinging lights connect the office area with the mainland of "Het Velp".
The relation with the office-area gets an extra meaning because of the light program.

目标
为办公区创造一个实时的艺术品。
灵感
作品的灵感来源于水面连绵的波浪和风中轻柔的摇摆。
它是什么？
它是一系列超过4米的空心易弯灯管。细小的直径确保它们能掀起一阵波浪舞。
工作原理
杆头可由电脑控制开关，并且灯光强度也是可调节的。
地点
摇曳的灯光将办公区和威尔普地区连接起来，给办公区与外界的关系增添了新意。

# The Light Orchestra

交互式艺术装置
——光乐团

**Credits**
Designers: Daniel Simonini (Italy/Brazil) with Lorenzo Marini (Italia) and Fernando Gonzales Sandino (Guatemala)
Master in Industrial Design
Supervision: Huub Ubbens
Production: LED contest with the support of APG Group

**Interactive Installation**

Light Orchestra is music for the eyes. It associates sounds with a perceptive domain different from hearing. The highlight of the installation is an oversized keyboard resembling a futuristic piano with 12 coloured buttons that control 12 LED sources. The visitor interacts with the instrument and creates changeable atmospheres by painting a vast, empty space with a palette of lights, colours and sounds. He is truly at the center stage, as the director of an imaginary orchestra. This grand piano whose shape is modelled on a waveform graph casually recorded, combines colours and sounds according to Goethe's colour wheel, based on 12 colours, and to the circle

of fifths, expressing the relationship between the 12 tones of the chromatic scale.

Light Orchestra took part in the LED Light Festival making his debut in December 2009 at the Colonne di San Lorenzo, a popular Milanese meeting place. It was voted as the best installation in the schools category and was awarded the second prize overall.

The new episode, Light Orchestra Volume 2, leaves the monumentality of the church's ancient colonnade and moves into a modern post-industrial cathedral, the former Faema factory.

作品"光乐团"的设计者是米兰工业设计学院(SPD)的学生。"光乐团"是视觉音乐，结合了不同于听觉里的主导元素的声音。这一装置的亮点体现在超大尺度的键盘，形似未来主义的钢琴，设计了12个彩色按钮，用于控制12个LED灯。参观者与这一乐器互动，运用灯光、色彩和声音的不同配置，着色于这一空旷、广袤的空间里，创造富于变幻的氛围。参观者真正地处在舞台中央，成为虚拟乐队的指挥家。这个庞大的钢琴，呈波浪状，以歌德的旋转色轮、12种颜色、五度音环为基础融合了色彩和声音，表现出半音音阶的12个音级之间的关系。

"光乐团"于2009年12月在LED灯饰节初次亮相，灯饰节的举办地为圣劳伦佐柱广场这个米兰人常去的聚会场所。这件作品被公众投票推为校园类最佳装置，进而获得了灯饰节二等奖。

新的乐章"光乐团2"离开圣劳伦佐教堂的古老柱廊，搬进了一个在Faema工厂旧址上建造的现代后工业风格的天主教堂。

# Alto Calore Aqueduct

## 深蓝里的黑色影像
——高山上的"水之萨满"雕塑

**Credits**
Location: Monte Pizzuto, Solopaca (BN) – Italy
Lighting Design: Filippo Cannata
Lighting Design Team: N. Fiorillo
Contracting Authority: Alto Calore Servizi Spa
Artist: Mimmo Paladino
Intervention: Interior and Exterior Lighting
Photography: Pasquale Palmieri

In the immediate proximities of Solopaca, a little town near Benevento, in the south of Italy, on the Pizzuto mountain, you will discover a place where you can spend some romantic moments in contact with nature. This is the new public square built in front of the city water system.

The water reservoir was created within the Regional Park of Taburno and Camposauro which covers 12,370 hectares and houses a population of around 25,000 inhabitants. Born for the protection of the massif Taburno Camposauro which is part of the Campanian Appennines, the Park offers valuable natural resources and landscapes in a context of great historical and cultural interest.

The entrance of the place is a work of art. The artist (Mimmo Paladino, the famous exponent of the Transavanguard Movement) healed with a blue dye the functional cuts that the engineer has inflicted on the mountain and a heart covered with a cobalt blue grit opens in the rock.

Light with its soft glows and the sound of the water fills the empty blue.

For this project Filippo Cannata and his partners faced a number of issues. Firstly, the location of the place: the site lies in the mountains, about 600 meters above sea level, where light pollution is zero and where the charm of the moon and stars at sunset reigns supreme; then the preservation of species classified as protected animals: WWF reported that the place is just on the path of migration of particular species of endangered birds; finally the interaction with the other people involved in the project: they had to interface not simply with an architect, but with an artist like Mimmo Paladino, known in Italy and abroad.

The use of non-invasive technologies enabled these lighting designers to achieve significant results and finally they won the challenge.

The built architecture is lacking of decorations, minimalist, industrial, planned by spatial relationships and imperceptible guidelines, in a great explosion of reflexes which are emanated by the glass shot, sometimes by the glassy blocks or by the water. Concrete is the only material used for the construction of the tunnel, even for the outside: the squares, the fountain, the walkways are all realized in a single vision of grey that is in harmony with the blue of the big wall. The design of the concrete structures is conceived in order to highlight the large stone blocks that interlock with one another, the attention to detail is superfluous, but the important is the grain of the material. All the geometric context leads to a single centre, the sculpture of "the shaman of water", a dark shadow in the deep blue.

The project is mainly characterized by the strong will not to interfere with the surrounding environment preserving the beauty of the effects and reflections generated by natural light. Any form of amplification or special effect was carefully avoided, the nature has the role of protagonist while the artificial light has a secondary, accessory function. The light is conceived as a pleasant background enlightenment, connected in sync with the slow cycles of the moon. The two different moments of lighting, contraction (systole) that is background light and relaxation (diastole) that is more intense light, are closely related to the four lunar phases: full moon – minimal lighting, no moon – brightest light. And the two variations in between. But even in the stages of higher intensity light is never excessive, never requires offending eyes, the lighting level is always contained, measured, in order to preserve intact the balance with the natural ingredients of the place and generate a steady suffused atmosphere of great emotional impact. The intensity of light in the different areas is carefully adjusted leaving some in the shade; special dichroic filters allow to get the right colour of light able to emphasize the architectural materials and elements of nature avoiding any form of glare.

The frequency and rhythm of the earth align themselves with those of the moon and the work draws its charm, its artistic value, its own meaning from this relationship, from this dialogue between artificial light and natural light that alternates sequences of peaceful tones to modulations and more definite expressions of the one and the other interlocutor. During the full moon dialogue softens, becomes more intense and the exchange reaches the highest level of intimacy: the artificial light, gentle, condescending, minimal, restrict itself to exploit the reflection in the water of the natural one. The employment of fibers optic and LED sources allowed us to achieve two fundamental objectives, a significant reduction of energy consumption and an equally significant reduction in costs. The project is in line with the principle of environmental sustainability and with a new vision of the future that has completely changed the criteria and rules of design. The employed LED light sources and electronic components provide the highest levels of durability; the significant reduction of power consumption, which is the most important goal, produces a huge extension of device's maintenance times with the consequent elimination of burdens and problems related to disposal. Moreover the devices are discrete, absolutely unobtrusive, able to camouflage in the architecture hiding from view and avoiding any possible forms of tampering or damage.

The realized interventions does not interfere with the brightness of the sky, rather it draws it in a primitive dance made of light and water. The visitor does not have barriers but views to see or glimpse, parking areas which consent to watch everything, a few moments to reflect on an example of restoration of the countless wounds of the Italian landscape.

意大利南部贝内文托附近的一个小镇索洛帕卡，在其最靠近的地方，就在皮兹托山上，有一个能接触自然度过一些浪漫时刻的地方，这就是在城市水资源系统前方新建的公共广场。

水库所在的塔布诺-坎波绍罗山区的区级公园，总面积12 370公顷，聚居了大约25 000位居民。为保护塔布诺-坎波绍罗山区，即坎帕尼亚-亚平宁山脉的断层块而建，具有深厚历史与文化背景的公园提供了宝贵的自然资源和景观。

入口区域俨然是艺术作品。艺术家(米莫·帕拉迪诺，意大利超前卫艺术运动的著名支持者）用蓝色染料修复了一个工程师在山上留下的功能性切口，并以中蓝色碎石填满了岩石里的心形缺口。柔和的灯光和水声充实了蓝色的空寂。

费立波·卡纳塔和他的合作伙伴在设计这个项目的时候，遇到了许多难题。首先，场所的区位：基地坐落在山里，海拔600米，此地的光污染为0，日落时的月色和星光最为迷人；其次，物种的保护：据WWF报道，此地为濒临灭绝的鸟类等特殊物种的迁徙之路；最后，与项目设计其他相关人员的互动：他们不仅只是简单地和建筑师合作，还包括与诸如米莫·帕拉迪诺这样一个意大利国内外知名的艺术家合作。

非侵入性技术令这些照明设计师取得了巨大的成果，并最终战胜了挑战。

建筑物缺少装饰，潜移默化中空间关系呈现极简主义工业风格。混凝土是建造隧道的唯一材料，甚至是在外部：广场、喷泉、步道都体现出单一的灰色，这与大面墙体的蓝色相和谐。混凝土结构的设计旨在突显那些块块相扣的大石块，细节的关注显得多余，但是材质细部却十分重要。所有的几何背景都导向一个单一的中心，即"水之萨满"雕塑，深蓝色里的黑色影像。

项目设计的特色在于不破坏周围环境的强烈意愿，力图保持自然光线反射的美学效果。任何形式的扩大和特效都是要小心避免的，自然是主

角,人造光是配角,起辅助功能。光作为令人愉悦的背景,与月亮的缓慢周期保持同步。照明的两个迥异时刻——收缩和舒张,前者为背景光,后者为强光,它们都与四个月相紧密相关:满月时照明打得最少,无月亮时照明打到最亮。其他月相则介于之间。但是,即使是最强的照明,也不会过度,旨在不伤害人的眼睛。照明等级总是处在可控范围内,力图保护自然元素的平衡与完整,所产生的稳定而曼妙的氛围能激发人们美好的情绪。

不同地区设置不同的光强度,有些保持阴暗;特殊分色滤片能获得所需的光色,进而突显建筑材料和自然元素,避免任何形式的眩目的光。

大地的频率和律动与月亮保持一致,照明设计则从这一关系中激发了它的魅力、它的艺术价值、它自身的意义,通过人造光和自然光的对话,调节了此地的平和度,使其更加彰显对话的张力。满月时,对话更加柔和,更加深切,交流达到了亲密无间:人造光柔和、谦逊,约束自己以利用自然光在水中的反射。

光纤和LED灯的使用,符合了两个基本目标,即大量地降低能耗和同等程度地减少成本。项目关于环境的可持续性原则和对未来的愿景,早已全方位地改变了设计标准和法则。LED灯和电子构件提供了最高等级的耐久性;能耗的大幅降低这一最为重要的目标,大大延迟了设备维护的时间,也必然消除了与弃置有关的负担和问题。而且这些设备分散装置,完全隐形,能够隐藏在建筑物里,不为人所看到,避免了人为的损害和毁坏。

建成的介入物并没有破坏天空的亮度,反之,它却以一种原始的光和水的舞蹈为环境增添一丝灵动。

# East Fremont Street

霓虹游乐场
——弗里蒙特东街街区

**Credits**
Location: Las Vegas, Nevada, USA
Client: Las Vegas Redevelopment Agency
Office: Selbert Perkins Design (SPD)
Designer: Selbert Perkins Design Robin Perkins, partner in charge John Lutz, principal in charge Andy Davy, design director
Affiliates: Civil Engineer: Kimley-Horn Civil Engineers
Frabricator: Fluoresco Lighting & Signs
Photography: Eric Staudenmaier

Selbert Perkins Design, hired by the City of Las Vegas, took cues from 1950s Las Vegas to create a visual feast of retro-futurist environmental graphics and street sculptures that embody Fremont East. Acknowledging that Las Vegas is known as the "neon playground," the designers intended to tell as much of the story of Fremont East during the day as the neon lights depict throughout the night, and therefore designing the streetscape to be engaging around the clock.

The entry gateway presents neon letters spread across an aluminum arc-shaped truss that is supported at each end by a conical stainless-steel pole. Adorning each pole is a painted-aluminum stylized arrow and starburst. The 24-point starburst dazzles from atop the support poles, while the arrow, with neon tube lighting, resembles in shape the entry sign's boomerang letter "F."

Four neon sculptures — a martini glass, retro "Vegas" sign, showgirl, and red stiletto — command the median with a flair that reinforces the district's playful tone.

These painted aluminum shapes soar atop support poles at heights of 15 to 42 feet, and all are illuminated by combinations of surface-mounted lights: neon, LED, and incandescent. The "Vegas" sign incorporates the stylized arrow and starburst features of the entry gateway, while geometric elements of the martini glass and dancing girl sculptures further emphasize the retro theme.

Existing light poles lining the streets

throughout Fremont East are glamorized with identity signs that celebrate the new district's identity. Seven poles — alternating from the north to south sides of the three-block strip — are adorned with a painted aluminum sign base that slightly curves at the bottom. The street-facing sign vertically displays the word "Fremont" in exposed neon block letters, while "East" is inset beneath horizontally in exposed neon script to brand the street.

Cast-bronze pavement medallions describing notable dates in Las Vegas history pepper the district's walkways. Diamond-shaped medallions adorn the crosswalk, while several oval-shaped ones scatter the sidewalks, totaling 18 in all. The medallions feature raised type and graphics, as well as etched patterns. In order to maintain consistency throughout the area, Selbert Perkins Design collaborated with the City of Las Vegas to develop guidelines for storefront signage for new business owners entering the district. Many existing businesses opted to comply in order to uphold the improved atmosphere, established by the design firm.

受拉斯维加斯市政府的委托，Selbert Perkins Design设计公司承接了弗里蒙特东街的改造工程项目，拟建一条以20世纪50年代的拉斯维加斯城市风格为背景的主题街道。街景由各式雕塑点缀，向市民和游客们展示复古未来主义氛围之下的一场视觉盛宴。拉斯维加斯素来有着"霓虹游乐场"的美誉，因此，景观设计师们计划在街景中添置大量的霓虹灯设施，将弗里蒙特东街的夜晚装扮得如白昼一般亮堂，并且这些景观灯将通宵为游客点放。

站立于街道的入口处，映入眼帘的便是极富曲线美感的"Fremont East District"艺术字霓虹灯，它依附于一个铝质弧形桁架之上，而这一桁架则由其两端的两根锥形不锈钢支柱支撑而起。这两根支柱的上端装有两个色彩鲜丽的铝质小箭头装饰物，柱顶还装点着两颗光芒四射的星状物。这两颗星状物在阳光和霓虹灯光的照射之下，显得格外耀眼，令人眼花缭乱；而紧随其下的两个小箭头装饰物，在霓虹灯管的包裹之下，闪闪发亮，与"Fremont East District"艺术字霓虹灯之中的字母"F"相互应衬着。

进入弗里蒙特东街街区，四个引人注目的霓虹灯雕塑便径直呈现于游客面前——巨型的马提尼杯、复古型"Vegas"字样灯箱、一个风情万种的广告女郎和一只高贵而性感的红色高跟巨鞋。每当夜幕降临，在这四个巨型霓虹灯的装点下，整个街区沉浸于一片灯红酒绿之中，弗里蒙特东街变成了拉斯维加斯城中的一个"不夜城"。

这些闪烁着耀眼光芒的各式铝质霓虹灯雕塑都悬挂在5至13米的高空之上，它们的表面都分别安装着霓虹灯、发光二极灯管和白炽灯等照明设施，使得它们在夜色中绽放异彩。复古型"Vegas"字样灯箱与巨型黄色铝质箭头饰物和柱顶光芒四射的星状物相互应衬，令街区的入口通道显得格外光彩夺目；而集结着众多几何元素的马提尼玻璃杯雕塑和体态婀娜广告女郎雕塑都突出了该街区的复古主题。

现有的这些霓虹灯柱贯穿于整个弗里蒙特东街区之中，为街景增添了几分迷人的色彩，也张显了改造之后新街区的动人魅力。在这一三叉路口区域内，7根霓虹灯柱交替着出现在人们的视野当中，在五光十色的铝质灯箱基座的反射之下，灯柱显得分外耀眼，而灯柱映射在基座上的倒影，若隐若现，给人以梦幻般的视觉享受。街区出口处的"Fremont East"霓虹灯之中，单词"Fremont"是纵向摆放的正楷字体，而"East"则是水平放置的手写体，这样独特的设计能够给来往的行人留下深刻的印象，从而，使得"Fremont East"渐渐深入人心，成为当地当之无愧的品牌街区。

街区内铺着青铜花纹路面的人行道，向过往的行人们诉说着拉斯维加斯曾经辉煌的历史。路面上18个形状各异且富有立体感的钻石与椭圆花纹相互搭衬，成为了街区内又一道亮丽的特色风景。为了保持整个弗里蒙特东街街区风格的一致，Selbert Perkins Design设计公司与拉斯维加斯当地政府联合制定相关准则，以便新进驻街区的商铺业主对其有所了解，并按照规定设计出与街区整体风格相符的店面和灯箱等。街区内现有的许多商铺业主都能够十分自觉地遵守有关规定，共同维护街区整体形象的统一。

# 05

# Thesis
# 先锋论点

■ ——Night Landscape
灯光与自然的和谐共处景观照明

A strategic view of lighting demands that the designer not only consider visual function but other design criteria such as creating an appropriate image, providing safety, improving the perception of security, assisting with way-finding or supporting the need to provide access for all.
从战略眼光来看，照明设计不仅要求设计师考虑到它的视觉功能，同时还有其他设计元素，如营造一个整体外观、提供安全设备、加强安全感知性、辅助路标或为所有进出口提供支持。

# Night Landscape

## 灯光与自然的和谐共处
——景观照明

**Credits**
Landscape Material by Lightem Gustavo Avil
Project Name:
Colima
Hacienda Jalmolonga
Hotel Encanto
Huatulco

A strategic view of lighting demands that the designer not only consider visual function but other design criteria such as creating an appropriate image, providing safety, improving the perception of security, assisting with way-finding or supporting the need to provide access for all.

There are a number of different approaches to masterplanning with light that have evolved in recent years. The most common is the development of frameworks for deploying public light in urban situations after dark. In most cases these are led by local government initiatives. The limits on such studies are that, despite light touching all parts of the visible city, the authority quite often only has control over basic public amenities such as streets, square and parks and a limited number of public buildings. This can greatly limit approach.

Recently other types of strategic lighting visions have a controlling interest in majors interventions of significant scale. In these cases not only the streetscape may be tackled but also the urban grain. This type of project also often has sufficient funding to implement ideas within a reasonable timescale, meaning visions can be fully realized.

These larger projects provide the opportunity to educate maintenance and management teams for urban enviroments about comfort, attractiveness and commercial success. Good lighting can encourage

people to inhabit spaces after dark that might otherwise be empty.

The final type of lighting masterplan is that for edge of town, or out of town. This is often adjacent to the green belt or in suburban or even rural locations. Here, consideration of the impact of the lighting requires a highly sensitive approach which seeks to minimize the impact of an artificial intervention withing a natural location; synthetic light in direct conflict with nature.

———Anne Bureau

Discuss the landscape is from my point of view a combination of elements that are visible in natural or urban context, as well as vegetables presences of natural elements.

Before darkness caused us fear, commanded respect, but over time this has changed, now incorporating more and more lighting we can enjoy the scenery, the outdoors at night has become an element of attraction.

This gives rise to the following reflection: Lighting can be applied in creating a night landscape, different from a day one, completely new and surprising, as it deletes and discover the visual information that we have on our environment.

It is not uncommon to see natural landscapes or urban open spaces illuminated the same way. But every landscape has to have its own illumination. Lighting is made with light but also with shadows. In other words, we have a long and difficult path, unique in many aspects, which implies the involvement of relevant institutions and society. A long way thus requiring a consistent and coherent discussion of the subjects involved. A path is unavoidable if you want to consider the landscape as a resource to preserve the memory and build the future.

从战略眼光来看，照明设计不仅要求设计师考虑到它的视觉功能，同时还有其他设计元素，如营造一个整体外观、提供安全设备、加强安全感知性、辅助标识导视或为所有进出口提供支持。

近年来形成了许多不同的照明规划手段。最常见的是给入夜后的城市环境排布公共照明设备。这种做法的局限性在于，即使照明设备遍布城市的每个角落，当局往往只控制了基本的公共区域，如街道、广场、公园，和一些公共大楼，这严重影响了效果。

其他一些战略照明大规模地采用了股份制。在这些案例中不仅街景得以保证，同时城市也成为受益者。这种类型的项目通常有充足的资金在合理的时间段里支持项目，这意味着可以完全实现规划中的效果。

大型项目在舒适性、吸引力和商业成功上给城市环境提供了培养维护与管理团队的条件。好的照明可以鼓励人们在夜晚使用原本闲置的空间。

最后一种照明规划类型是为城镇角落或乡村设计的。它通常毗邻绿化带或是郊区，甚至是乡下。这里的照明效果需要使用一种高度敏感的手段来将人造设备对自然界的影响降至最低，避免光线与自然界的直接碰撞。

——安妮·布鲁诺

以我的观点看,景观就是自然界或城市环境中包括植物等可见的元素的集合。

从前,夜晚使我们感到恐惧,乃至敬畏,但是时光荏苒,这已经改变了。如今越来越多的灯光使我们能在夜晚尽情享受周围的环境,夜晚的室外也变成了一个具有吸引力的元素。

这导致了以下的结果:灯光的应用可以制造出一个完全不同于白天的景观,崭新而惊人的,它消去和显露了我们环境里的视觉信息。

自然景观或城市开放空间采用相同的照明方式并不奇怪,但是每一处景观都应有自己的照明。照明是由光和影一起组成的。

换句话说,道路是曲折的,在实践照明设计的过程中会碰到许多棘手的问题,牵扯到一系列相关机构和社会因素。

路漫漫其修远兮,吾将上下而求索。如果想要将景观作为一种保存回忆、构筑未来的资源,那么披荆斩棘将是不可避免的。

# LED, Sustainability, Preservation and Art

## 保护与艺术
## ——LED灯的可持续性

**Credits**
Office: Barbara Balestreri Lighting Design
Photography: Al-Fann picture by balestreri studio
Arcimboldo picture by balestreri studio
Japan picture by balestreri studio
Lucio Fontana_Ceiling_picture by Daniele de Lonti
Monastery_Torba_Tower_picture by Boutique Creative

The intrinsic characteristics of the LEDs allow, especially in the lighting of artworks, to overcome the limitations of each type of sources, allowing viewers to return color and not subject to the total chromatic optical phenomenon linked to the interpretation of the color the human eye.

In recent years, the lighting design of artworks, we mainly have used the LED products by Osram because light sources are superior to traditional in many ways and are distinguished by their extreme versatility. These LEDs are striking for the huge range of colors and excellent color saturation. The low power consumption and long life make them particularly cheap, and above all they have a reduced environmental impact making the source a product with high value of sustainability.

With these characteristics, LEDs have become essential for our lighting design especially those relating to artistic and cultural significance. These LEDs are simple, efficient, free of emissions in ultraviolet band preserving the integrity of artwork and create more fascinating light effects.

The use of LED combined with the ability to adjust the intensity of the issue becomes of paramount importance, especially for those light-sensitive materials and must follow strict rules of conservation.

LED灯固有的特征能帮助艺术品照明设计突破各种光源的局限，使观者回归颜色，而非局限在与人眼有关的彩色光象里。

近年来，艺术品的照明设计主要采用了欧司朗牌的LED产品，因为LED灯的光源在很多方面都优于传统的灯源，其多样化最为引人注意。LED灯的颜色涵盖范围广，色彩饱和度出色。低能耗、寿命长，且十分便宜，最为重要的是LED灯对环境的影响十分小，具备高度可持续的价值。

LED灯因其有着上述特征而成为照明设计中不可或缺的元素，特别是那些与艺术和文化意义有关的设计。这些LED灯简单、有效，且不释放紫外线，能够保护艺术品的完整性，营造出更加迷人的光效。

LED灯的使用融入了灯光强弱的可调节性，特别是那些对光敏感的材料，这至关重要，还必须严格遵守保护条例。

## Al-Fann. Arts of the Islamic Civilization. The al-Sabah Collection, Kuwait

The exhibition Al-Fann. Arts of the Islamic civilization. The al-Sabah Collection, Kuwait which has seen us involved with the organizers of Skira Editore, with the curators Giovanni Curatola, Sue Kaoukji of Kuwait and with the designer of the exhibition, architect Corrado Anselmi, presented to the public works of various kinds: ceramics, glass, textile, miniatures, wood, metals, etc., which illustrate, through creations of art and craft, a culture from different geographical areas and historical periods through fifteen centuries.

It was necessary, in this work, a dedicated lighting for each artwork on showcase, considering the complexity of the materials exhibited extremely different from each other like very small objects such as coins and big carpets. The lighting design was integrated completely with the design of the exhibition.

The ingredients for the development of the lighting project were the high technology of light sources, sources with warm temperature of colour, a high colour rendering, anti UV, anti-IR, fixture and a light intensity adjustable. All these elements were of crucial importance especially for those light-sensitive materials, present in most exposure and therefore require small values of illumination.

The halogen light, warm light gave a more punctual light to the different showcase and on the set design while the LED light, also warm but more widespread and homogeneous, was internal to some showcase. This has enabled us to develop in different ways the objects and to maintain the lighting values always below the maximum permitted by the regulations. Integrate with architectural and setting context means, not only highlight the works to make them emerge from the context in which they appear, but also create special lighting suggestions that

emphasize the complexity of Islamic civilization, making it more attractive the exhibition path. The "file rouge" was create a warm atmosphere with soft lighting on the artwork and flashes of light that emphasize; the visitor discover the Islamic world, made of extremely fine details both in terms of decoration, colours, materials, manufacturing.

艺术·伊斯兰文明的艺术·科威特萨巴赫作品展

这个作品展集结了照明设计师、承办方Skira Editore出版社、科威特策展人乔瓦尼·库拉托拉和苏·卡奥奇、展览设计师、建筑师科拉多·安塞尔米，展出了种类丰富的公共作品，如陶瓷、玻璃、纺织、微型画、木材、金属等，通过艺术和工艺的创造，展现了横跨不同地理区域、纵越15个世纪的历史时期的文化。

考虑到展品材料的复杂性和相异性，如硬币等小件物品以及大幅地毯等，陈列柜上的每件艺术品都必须设计专门的照明效果。照明设计与展览设计完美地相整合。

照明元素包括光源的高技术、中等色温、高彩效果、防紫外线、防红外线、可调亮度的灯具。对于那些大部分暴露在日光下且只需要些微照明的光敏感材料而言，这些元素至关重要。

卤素灯的暖色调照明能给不同陈列柜和成套设计带来更加精确的光源，而LED灯虽然也是暖色调，却更为发散些，适用于一些的内部陈列柜。由此，照明设计师得以采用多种方式设计照明效果，使其总是处于规定许可范围内的最大值。

与建筑和背景环境相融合，意味着使展品从其所处的背景里呼之欲出，同时营造出特殊的照明来突显伊斯兰文明的复杂性，使展区过道更加吸引人。

"file rouge"充满温暖的氛围，柔软的灯光洒在艺术品之上，配以强化的闪光灯；游客在此探索伊斯兰世界，发现极致的华美细节，如装饰、色彩、材料、制造工艺等。

## Arcimboldo. Milanese Artist between Leonardo and Caravaggio

In the exhibition Arcimboldo. Milanese artist between Leonardo and Caravaggio next to the artist's famous paintings that depicting the four seasons and the four elements of Aristotelian cosmology in the form of an allegorical portrait, sketches and drawings are exhibited, including naturalistic studies, choreography for pageants and parties, portraits, and objects obtained by the combination of a variety of anthropomorphic forms that explain the path done by the artist to get to the creation of his composed heads and grotesque faces.

The desire to recreate a magical and a bit fantastic dimension, that emerges from the Arcimboldo's works, has guided the design of light on solutions in which, the light source hidden in the furniture or showcases, could reveal and emphasize the uniqueness of individual works, realized with different techniques and materials. The lighting project has a goals to reach a correct vision and perception of the artwork respecting single conservative needs. Use the dimmable system it allowed us to adjust the light intensity for each artwork.

The lighting design has led to some choices of interior design. Light has also influenced the interior design to give emphasis, create a light scenography, like the backlit show case.

On the great "composed heads", as well as other oil paintings along the exhibition way, the cutting and very effective lighting (halogen source with a narrow beam), emphasized the uniqueness of the work of the Milan

artist bringing out of context paintings of small size and the minute details of the great works. All the LED and the sources employed have a colour temperature of 3000K and they are powered and dimmered, appropriate to single conservative needs and of correct vision and perception; they allow to adjust each time the light intensity.

Projectors were used with different lenses (fresnel and diffusing) and with different accessories (honeycomb and flap). This is to ensure that the light could have a varied language.

The LED with its small dimensions, emissions-free in the infrared and ultraviolet spectrum has allowed the integration of the lighting in the setting, diminishing the distance from the work of art and underlining works that need low lighting level with sharp and incisive lines of light. It was possible illuminate each work of art, even those more delicate from a conservation point of view, putting in evidence every smallest detail that couldn't be properly revealed with a traditional source.

阿尔钦博托展,介于李奥纳多和卡拉瓦乔艺术风格之间的米兰艺术家

阿尔钦博托的著名画作以讽喻肖像的形式描绘了四季和亚里士多德宇宙论中的"四因说"。画作附近的"阿尔钦博托展"展出了素描和手绘图,包括自然主义研究,庆典和宴会的动作设计,肖像,以及以各种拟人的形式组合而成的物品,阐释了画家如何创作出他那些组合起来的头部和怪异的人脸。

再现阿尔钦博托作品中魔幻而略带荒诞的维度,这指导着照明设计的方案形成,其中,灯源隐匿在家具和陈列柜里,以此突显和强化单一作品的个性,同时配以不同的技术和材料。照明项目的目标在于形成一个对艺术品的正确表现和理解,尊重每件作品的保护需求。利用可调光系统达成调节每件艺术品光的亮度的目标。

照明设计引出了一些室内设计的手法。灯光也影响室内设计突出重点,创造出光透视法,如背光陈列柜。

在那些伟大的"组合头像"画,以及展览的其他油画上,切光和高效能照明(发出狭窄光线的卤素灯)突显了米兰艺术家作品的独特性,衬托出小尺寸的背景色和大幅画作的精微细节。所有的LED灯及其他光源的色温都设置在3 000开尔文,并配以电源和调光器,不仅满足了单件作品的保护需求,还正确地传达了对作品的表现和理。

投影机采用不同的镜片(棱镜和漫射镜),安装不同的配件(蜂巢式光闸和袋盖),以此确保灯光的不同效果。

LED灯体格小,不会发出红外线和紫外线的光,这些特性使LED灯得以被设为集成照明,拉近了和艺术品的距离,用光的鲜明而深刻的线条突现了需要低照明等级的艺术品。每件艺术品有专门的照明,可以展现出那些传统光源无法展现的每一个最微小的细节。

## Japan. Power and Splendor 1568 - 1868 Was an Exhibition

The exhibition Japan. Power and Splendor 1568 - 1868 was an exhibition that involved with the curator Prof. Gian Carlo Calza, the designer of the exhibition Roberto Peregalli and Laura Rimini architects, the organizers 24ORE Motta Culture, to the providers of works, especially came from Japan.
Junichiro Tanizaki's book, "In praise of Shadow", was the inspiration for the mood of the whole construction to achieve an atmosphere linked to Japanese culture.
Interface with fragile and delicate artworks, such as paper or silk screens, was a very delicate and careful work. Hence the choice to use, as a light source of the exhibits in the showcases, the dimmable LED.
The illumination of each showcase, in fact, can be regulated at the level of intensity. This allowed us to valorise the works on showcase in a different way and to maintain the lighting values always below the maximum permitted by the luxs Japanese regulations.
The choice, finally, an ambient light of twilight, with light controlled levels, interrupted by flashes of light, emphasizes the artworks on showcases and on tables, enhancing them and making them readable even in their nuances.

## 日本·1568年—1868年权力和荣耀展

照明设计师与各方进行合作,如策展人吉安·卡尔洛·卡尔萨教授,展览设计师罗伯托·佩内加里,建筑师劳拉·利米尼,承办方24ORE莫塔文化出版社,展品提供者,特别是来自日本的展品提供者。

整个照明构建的灵感都源自谷崎润一郎的《阴翳礼赞》,旨在营造一种与日本文化相关联的氛围。纤细精致的艺术品,如纸质或丝质屏风,与之有关的照明设计也是一项需要细致和小心翼翼的工作。因此,可调光LED灯运用于陈列柜成为最佳选择。每个陈列柜的照明亮度都可以调节。这使得设计师得以针对不同的陈列柜设置不同的亮度,使照度总是处于日本规定最大值的范围内。

由此,可控级别的微亮的环绕光,插入闪光,突出了展柜和桌子上的艺术品,不仅美化了艺术品,还突显了每一处细节,使其清晰可辨。

## Ceiling by Lucio Fontana

The restoration of the beautiful Ceiling by Lucio Fontana of 180 square meters and made on 1956 for Hotel Gulf of Procchio in Elba island and re-installed on the last floor of the Museo del Novecento of Milan, in the tower of Arengario, has provided, for reasons of preservation and maintenance, also the replacement of the previous light sources. Our intervention provided for the use single LED placed in the holes where Fontana had previously placed incandescent light sources.

To re-create the lighting effect desired by the Maestro Lucio Fontana, we made many tests with LED light sources with different power and temperature of colour and with glass with various finishes to close the sources and diffuse the light. In agreement with the restorer, Dr. Barbara Ferriani, we selected the LED Osram, specifically DRAGONpuck, with a definite power and temperature of colour of 2700 °K.

The result was that of a soft warm light, diffused that has completed a careful restoration of one of the greatest works of art created by Maestro Lucio Fontana.

### 卢西奥·丰塔纳的天花板作品

这个美丽的天花板总面积为180平方米，由卢西奥·丰塔纳建于1956年，最初安装在厄尔巴岛的普罗吉奥海湾酒店，之后出于保护和维修则移装至米兰20世纪博物馆的最后一层，即阿刃噶里奥塔的旧址，为了保存和维护而进行了整修，同时更换了之前的照明光源。此项目的照明设计师移除了丰塔纳放在洞里的白炽灯，换之以LED灯。为了重新创造卢西奥·丰塔纳想要的照明效果，设计师进行了各种测试，如对LED灯不同电源和色温的测试以及不同玻璃铺装所产生的遮光和漫射作用的测试。设计师与天花板修复员芭芭拉·费里亚尼博士达成一致，选择了欧司朗牌的LED灯，特别是DRAGONpuck这个高亮度LED产品，电源恒定，色温为2 700开尔文。

照明效果体现为轻柔的暖色调灯光笼罩着细心修复后的卢西奥·丰塔纳的伟大艺术品。

## Tower of the Monastery of Torba

In the conservation project of the Tower of the Monastery of Torba (Castelseprio Gornate Olona, Italy - FAI - Italian Environmental Fund), the light has an important role because, confronted with demanding regulatory and architectural especially the frescoes, is responsible of enhancing the architectural details, the environments and the surface that the time made precious. Very strong spiritual and mystical atmosphere embraces the military architecture of the tower and the interior of the two rooms.
The small size of LED source allowed the integration of the LED in to the architecture itself, hidden below the platform, lighting softly with a warm tone. The result was a light that illuminates from the bottom up all the walls in a uniform way without creating glare. Moreover, the choice of positioning the platform about 70cm from the painted walls has created a "natural" bollard that can limit direct contact with fresco, without additional technical elements too invasive.
The various lighting design choices have been defined and shared with the architect Corrado Anselmi and together with the FAI, the Italian Environment Fund.

## 托尔巴修道院的军事塔

在这个保护项目里，光扮演了十分重要的角色，其作用不仅体现在满足建筑的需求，特别是壁画，还体现在美化那些经过时间的洗练而更加珍贵的建筑细部、环境以及表面。非常强烈的精神和神秘氛围笼罩着整个军事塔建筑以及两个房间的内部。
LED灯的小身型使其隐入平台，与建筑相融合，散发出柔和而温暖的光。光效从底部沿着墙面向上延伸，整齐有序，没有丝毫眩目感。同时，平台距离壁画大约70厘米，营造了天然的护柱，进而减少了与壁画的直接接触，在此就无需介入其他入侵性的技术元素了。
各种照明设计的手法都是经过与建筑师科拉多·安塞尔米和意大利环境基金会(FAI)讨论后而决定采用的。

## 图书在版编目（CIP）数据

照明设计：建筑·景观·艺术 / 凤凰空间·上海编译. -- 南京：江苏人民出版社，2012.3
（最新照明设计）
ISBN 978-7-214-08027-1

Ⅰ.①照… Ⅱ.①凤… Ⅲ.①建筑设计：照明设计－案例－汇编－世界 Ⅳ.①TU113.6

中国版本图书馆CIP数据核字(2012)第045564号

## 照明设计——建筑·景观·艺术

凤凰空间·上海　编

| | |
|---|---|
| 策划编辑： | 冯 林　陈兰真　石 莹　田俊淼 |
| 责任编辑： | 刘 焱　潘 华 |
| 责任监印： | 彭李君 |
| 出版发行： | 凤凰出版传媒集团 |
| | 凤凰出版传媒股份有限公司 |
| | 江苏人民出版社 |
| | 天津凤凰空间文化传媒有限公司 |
| 销售电话： | 022-87893668 |
| 网　　址： | http://www.ifengspace.com |
| 集团地址： | 凤凰出版传媒集团（南京湖南路1号A楼 邮编：210009） |
| 经　　销： | 全国新华书店 |
| 印　　刷： | 深圳当纳利印刷有限公司 |
| 开　　本： | 965毫米X1194毫米 1/16 |
| 印　　张： | 27 |
| 字　　数： | 216千字 |
| 版　　次： | 2012年5月第1版 |
| 印　　次： | 2012年5月第1次印刷 |
| 书　　号： | ISBN 978-7-214-08027-1 |
| 定　　价： | 398.00元 |

（本书若有印装质量问题，请向发行公司调换）